T0292341

FIRST PRINCIPLES OF CHEMISTRY

FIRST PRINCIPLES OF CHEMISTRY

by

F. W. DOOTSON, M.A., Sc.D., F.I.C.
University Lecturer and Demonstrator in Chemistry, Cambridge

and

A. J. BERRY, M.A.
Fellow of Downing College
University Demonstrator in Chemistry
Cambridge

CAMBRIDGE

AT THE UNIVERSITY PRESS

1927

PREFACE

From a variety of causes, a considerable number of students practically begin the study of chemistry at the University. As this number shews no sign of diminishing, we have carried out a long contemplated project by producing a textbook more particularly designed for the use of such students; a book to be studied as an adjunct to lecture-room and laboratory teaching rather than one to be merely read. We think it should minimize the time taken up by writing notes in class—a practice which withdraws attention from what is being said or demonstrated by the lecturer.

The scope of the book amply covers the requirements of the School Certificate Examinations of all the English Universities, and we hope it may be found useful and acceptable for School teaching.

The order in which the subject matter is treated has been found in practice to work very well, and it certainly seems to us to be more logical than the older systems of textbook arrangement. Certain sections may be omitted with advantage on a first reading, at least by those who are really beginning the study of the subject, but these have not in general been indicated in the text. Individual students differ so much in ability and in requirements that selection is better made by the lecturer.

General principles and general methods have been emphasized throughout, and details have only been stressed where they have a special significance.

We do not consider the Periodic Law suitable for inclusion in an elementary course, since it can be studied profitably only in the light of a fairly extended knowledge of descriptive chemistry, and it has therefore received but little attention, while the now obsolete Laws of Multiple and Reciprocal Proportions have been omitted altogether. On the other hand, hydrogen-ion concentration and the chemistry of colloids are too important nowadays to be ignored even in elementary courses, while a knowledge of them is especially necessary to medical students, who preponderate in the first-year classes at the University. We do not find that they present exceptional difficulties to the average pupil, and a few pages dealing with these subjects have therefore been included.

In conclusion, we desire to offer our grateful thanks to our friends and colleagues C. T. Heycock, F.R.S., Fellow of King's College, Goldsmiths' Reader in Metallurgy, for his kindly interest and advice while the book was in preparation; H. McCombie, D.S.O., M.C., M.A., D.Sc., Fellow of King's College, University Lecturer in Chemistry, who read the book in manuscript, for his assistance and suggestions; and U. R. Evans, M.A., for his welcome help and criticisms of Chapter XI.

F. W. D.
A. J. B.

University Chemical Laboratory
Cambridge

January 1927

CONTENTS

CONTENTS

CHAPTER I

INTRODUCTION

IF marble be crushed to a fine powder, each of the particles possesses all the properties of the original substance except size and shape. This subdivision may obviously be continued so long as the particles are visible under the microscope, but it is difficult to believe that it must stop there. Is there a limit to the process of division? However small the particle may be, one can conceive of its being broken once more.

The extreme subdivisibility of matter can be shewn in a striking manner by the following experiment.

One gramme of fluorescein is dissolved in a solution of caustic soda and made up to a volume of one litre with distilled water (solution A). From this, 10 c.c. are withdrawn and diluted to one litre (solution B). The process is repeated with solution B and solution C obtained. The solution A contains 0·001 grm. of fluorescein per c.c. Solution B contains one-hundredth part of this, that is 10^{-5} grm. per c.c., and solution C, 10^{-7} grm. per c.c. If 1 c.c. of solution C be placed in a glass tube of small bore and held in a position oblique to the line of vision the fluorescence can be clearly seen. That is, there is optical evidence that 1 grm. of fluorescein is divisible into ten million parts, and there is still no evidence that the limit of divisibility has been reached.

If a piece of marble be heated strongly a gas is evolved and something remains behind which is not marble. This latter material therefore consists of at least two substances and if so, subdivision of marble must at last reach a point where the particles are *not* all alike in properties. A *chemical* separation has been brought about and two substances, having different properties, obtained.

This process of splitting up substances into two or more other substances of different properties, applied to all material things with which we are acquainted, has resulted in our obtaining some ninety kinds of matter which cannot at present be further resolved. That is, no one of them can be made to yield *two* substances of different properties without the addition of something else, which in practice means without an increase in weight. These substances are the chemist's *elements**.

According to Boyle (1661) 'elements are the practical limits of chemical analysis, or are substances incapable of decomposition by any means with which we are at present acquainted.'

For a century after Boyle hardly more than a score of so-called elements were known. Soda, potash and quicklime were 'elements' until early in the nineteenth century, when Davy obtained the metals sodium, potassium and calcium from them.

The term 'element,' then, is provisional. An element of to-day may to-morrow prove to be a compound, or even a mixture. Copper is *at present* an element because nothing has been obtained from it hitherto but copper, unless by the addition of something else and a consequent increase of mass.

It must not be thought, however, that the elements of to-day are *generally* open to suspicion. In some few cases perhaps they may be, but there is evidence from many different and unconnected sources that nearly all of them are incapable of being decomposed in the sense indicated above. This point will be again referred to later. The

* Though very nearly so, this is not *absolutely* true. See radium, p. 10, and allotropy, p. 73.

usual classification of elements into metals and non-metals has a rough convenience, but it is neither rigid nor important. The element arsenic shares the properties of both classes. Metals in general have what is known as a metallic lustre, some of them in a high degree, as palladium, but the non-metal iodine also has a quasi-metallic lustre.

It will be seen later that the non-metals as a class readily unite with hydrogen, forming stable compounds with it, while the metals as a class either do not unite with hydrogen at all, or do so with difficulty, forming far less stable compounds. Every metal forms at least one basic oxide, while non-metals do not form basic oxides. The metals are all good conductors of heat and electricity, while non-metals as a class are bad conductors.

If a mixture of the elements iron and sulphur be made by using the finest iron filings and 'flowers' of sulphur, however intimate the mixture may be it is still possible, under the microscope, to detect the particles of iron and of sulphur lying side by side apparently unchanged. Further, the iron can be withdrawn from the mixture by means of a magnet, or the sulphur can be withdrawn by shaking the mixture with a solvent such as carbon bisulphide, filtering, and allowing the clear liquid to evaporate from a clock glass, when the sulphur is left behind. The mixture of iron and sulphur is stable and permanent if it be kept cool and dry. If the mixture be heated in a hard glass tube a point is soon reached at which, the external heating being stopped, heat is generated in the tube itself. A change takes place in which considerable energy is evolved in the form of heat.

If the contents of the tube be cooled and examined, a substance is found which is neither iron nor sulphur; a new body in fact, which is not attracted by a magnet, which yields nothing to carbon bisulphide and which, when placed in dilute sulphuric acid, gives a gas with an odour quite unlike the one obtained by similarly treating the mixture. A chemical union has been effected.

The new substance is much harder than iron and much more brittle. It is sometimes used in a modified form for cementing the joints in the iron mains conveying gas or water. The iron-sulphur mixture, together with a little ammonium chloride, is rammed tightly into the socket between the pipes when they are in position. The moisture of the earth which covers them assists the change described above, in which the metal of the pipes themselves takes part, and after a time the iron surfaces are so firmly welded that it is often easier to break the pipes than to disconnect the joint.

A marked feature of this change is that the new substance on examination proves to consist always of 63·7 per cent. of iron and 36·3 per cent. of sulphur. If the original mixture did not contain the elements in these proportions, then the new body *either* contains unchanged iron and will still be attracted by a magnet, *or* some sulphur will be boiled off and will condense in the cold part of the tube.

This constancy of composition is more conveniently shewn in the case of common salt, since here the whole of the experimental work can easily be carried out by the student himself.

If a small piece of the element sodium be placed in a crucible in a large jar of the elementary gas chlorine, the sodium combines with the chlorine with the formation of salt. This can be collected, washed with a little concentrated hydrochloric acid and dried (specimen 1).

If sea-water be evaporated down to about one-tenth of its original volume, filtered to remove suspended impurities and then made strongly acid with pure concentrated hydrochloric acid, common salt is deposited from the solution. This can be filtered, washed with a few drops of water and dried (specimen 2).

If a solution of washing soda be slowly acidified with hydrochloric acid and evaporated to small volume, common salt is deposited and can be treated as above (specimen 3).

The percentage of chlorine (and therefore by difference the percentage of sodium) can be determined by analysis, when the same values will be obtained from all three specimens.

It may thus be shewn experimentally that (pure) common salt however obtained (and the examples given do not of course exhaust the possible methods of obtaining it) always consists of 60·7 per cent. of chlorine and 39·3 per cent. of sodium.

Another set of experiments shewing this constancy of composition may be made with black copper oxide.

1. If a little thin foil of pure copper be carefully weighed into a crucible and heated to redness with free access of air until the weight no longer increases, the masses of copper and oxygen which unite can be calculated.

2. If copper oxide be prepared by heating copper carbonate to a red heat; a known weight of this oxide taken and heated in a current of hydrogen until no further loss of weight occurs; the mass of copper remaining can be weighed and this, by subtraction, gives the mass of oxygen united with it in the specimen taken.

3. If a few grammes of copper nitrate be heated in a crucible, gently at first but finally to a red heat, a third specimen of copper oxide will be obtained which can be examined as in 2.

In the three cases the percentage of copper in the copper oxide proves to be the same, viz. copper, 79·9 per cent.; oxygen, 20·1 per cent. As in the case of common salt, the three specimens have been prepared by methods independent of each other, yet they are identical in composition.

The results of these and other experiments are summarized in the Law of Constant Composition (or Law of Constant Proportions) first clearly formulated as the result of a controversy between Berthollet and Proust (1803–1806). The former maintained that the composition of compounds might be dependent on the conditions of their preparation, but the careful experimental work of the latter shewed that compounds are of constant composition. The law is usually stated thus:

The same compound always consists of the same elements united in the same proportions.

This law therefore is to be regarded as nothing more than a summary of experience. Innumerable experiments confirm it, and hitherto no exception to it has been established.

Conservation of Mass and Energy. If a piece of dry phosphorus be placed in a dry flask which is then closed with a tightly fitting rubber stopper, the whole carefully weighed and then dipped for a moment into hot water, an obvious chemical change begins. The flask should be gently rotated while the phosphorus is burning to avoid danger of breakage. When the action is over and the flask has returned to room temperature it is dried and again weighed. No change of weight is observed. If the flask be opened under water a quantity of water enters, approximating to one-fifth of the volume of the flask if sufficient phosphorus has been used. The water which previously was without action on litmus, now turns it red. If too much phosphorus is used a portion of it remains unburnt.

The experiment shews that this chemical change takes place without loss or gain of weight in the system as a whole. Is it true of all chemical changes?

The principle of Conservation of Mass was first clearly stated by Lavoisier in the year 1789, as the result of experiments which he carried out on the transformation of sugar into alcohol and carbonic acid gas under the influence of yeast. His experiments on the calcination of metals in closed vessels in 1774 had previously indicated its truth. These experiments were essentially similar to that on the combustion of phosphorus just described. All subsequent quantitative experimental work has led to conclusions which are in agreement with the principle, but it is important to note that there is no *a priori* reason why it should be true.

The fundamental importance of the principle of conservation of mass led to very careful work in recent years to test its degree of accuracy. The most searching experiments are those of Landolt, who examined some fifteen different reactions. In his final experiments, published in 1908, the total reacting mass employed was about 400 grm. and the maximum error was 0·03 mg. Landolt concluded that the law of the conservation of mass can be considered as proved within the limit of error of one part in ten million. It is sometimes called the law of the indestructibility of matter, the meaning of which is, simply, that in any chemical change the total mass of the substances which interact is (within the limits mentioned) equal to the total mass of the products of the reaction*.

When the chemical change takes place spontaneously there is always a liberation of energy of some kind. Since heat is the lowest form of energy, the chemical energy

* More recently (Rutherford, 1922) the validity of this law, on which all the quantitative work of the chemist depends, has assumed a new aspect. 'We now know with certainty that four neutral hydrogen atoms weigh appreciably more than one helium atom, though they contain the same units, four protons and four electrons.' Aston, 1924.

liberated by the reaction is usually manifested (and measured) as heat. This is a particular case of the general principle of the conservation of energy, which was clearly established by Joule in 1850 as the result of his work on the mechanical equivalent of heat. The law of the conservation of energy may be stated in the following terms. In an isolated system the sum of the various kinds of energy is constant. If a quantity of one form of energy disappears, an equal quantity of some other form of energy makes its appearance. The statement amounts to this: if A unites with B to form C with the evolution of n units of energy, then n units of energy will be required to reverse the change and reproduce A and B from C.

Atoms and Molecules. We have seen above that the divisibility of matter is not infinite. A point is at length reached when further subdivision must result in producing substances with different properties. The smallest particle of marble which shews all the properties of marble is the *molecule*, a term which is defined as the smallest particle of matter which is capable of an independent existence. Subdivision of the molecule of marble produces smaller molecules of substances which are not marble.

An indication of the order of magnitude of these ultimate particles has been obtained in the following way.

A fragment of camphor placed on the surface of water appears to be in rapid motion, doubtless due to local alterations of surface tension, but a thin film of oil on the water prevents this movement. In a series of experiments Lord Rayleigh (1890) determined the minimal thickness of the layer necessary to stop the motion, which he found to be $1 \cdot 6 \times 10^{-7}$ cm. Now it cannot be assumed that the oil layer was only one molecule thick, but it cannot have been *less* than one, therefore this value represents the superior limit of molecular dimension, which is to say that the diameter of a molecule of the oil cannot be greater than sixteen millionths of a millimetre.

Lord Kelvin's estimate (1870) places the superior limit of molecular dimension at 10^{-8} cm. and the inferior at 5×10^{-10} cm. His illustration is given in the following terms. 'Imagine a raindrop, or a globe of glass as large as a pea, to be magnified up to the size of the earth, each constituent molecule being magnified in the same proportion. The magnified structure would be coarser grained than a heap of small shot, but probably less coarse grained than a heap of cricket balls.' More recent estimates assign somewhat greater values than these.

It has been shewn that common salt is a compound of the elements sodium and chlorine. The smallest particle of common salt which can have an independent existence, the molecule of common salt, is therefore capable of further subdivision; but this subdivision results in the realization of a second and smaller type of ultimate particle, namely, the smallest particle of an element that can take part in a chemical change. This latter kind of ultimate particle is termed the *atom*.

The conception of an atomic structure of matter dates back to very early times. It was revived by Dalton in his *New System of Chemical Philosophy* published in 1808. Dalton however did not draw a clear distinction between the ultimate particle of an element and the ultimate particle of a compound, which latter he spoke of as a 'compound atom.' In other words, Dalton's ideas did not extend to the modern conception of the fundamental difference between the molecule and the atom. The necessity of drawing such a distinction was first recognized by Avogadro who in 1811, introduced and defined the idea of the molecule in order to explain certain experimental results observed by Gay-Lussac on the combination of gases by volume. (See Chap. II.) We cannot therefore speak of an atom of a compound such as marble. The term would be meaningless, since the smallest particle of marble which is capable of a separate existence consists of three different kinds of atoms.

The molecule and the atom are identical in some few cases. Argon and helium, for example, have mon-atomic molecules, but as is shewn later, the molecules of nitrogen, chlorine, hydrogen, etc., consist each of two atoms firmly united in some way which we do not as yet clearly understand.

Up to the end of the nineteenth century the atom was considered to be indivisible and unalterable, and even at the present time it can still be considered as such, so far as chemical reactions are concerned. The existence of sub-atomic particles known as electrons was however placed beyond doubt by Sir J. J. Thomson (1897). Electrons, which can be considered as 'atoms' of negative electricity, are universal constituents of all atoms. Further, elements which are radio-active have unstable atoms, the most notable of which is the element radium. Ramsay and Soddy (1904) shewed that the atom of the element radium undergoes spontaneous disintegration, giving rise to atoms of helium and other elements. Atomic disintegration is a process which is independent of experimental conditions, and takes place at a definite rate which is fixed for each unstable element; that is, the rate of breaking down can be neither hastened nor retarded by any method known to us. Still more recently Sir E. Rutherford (1922) has been successful in artificially disintegrating in a minute degree the atoms of certain elements, notably those of nitrogen, into atoms of helium and hydrogen.

A characteristic property of atoms as formerly understood was the absolute identity of all those of the same element. The atoms of different elements, however, do in general differ considerably in mass. For the purpose of comparing the relative masses, the lightest atom, that of hydrogen, was chosen as the standard, and the atomic weights of the rest of the elements were expressed numerically in terms of it. This standard was subsequently abandoned in favour of a more convenient one, oxygen, the atomic weight of which was arbitrarily fixed at 16. This is the modern standard, under which the value of the atomic weight of hydrogen is 1·008. There is no doubt that the atoms of some of the elements are all precisely alike; other elements, however, consist of atoms which differ in mass. Although there is only one element chlorine, there is more than one kind of chlorine atom. These different kinds of atoms of the same element are termed *isotopes*. Isotopic atoms are identical in chemical properties; they differ in mass, and this is their only known difference.

One of the isotopes of chlorine has a mass of 35, while another has a mass of 37. These isotopes have never been completely separated from each other, and whenever and however chlorine gas is prepared, the isotopes are always found in it in the same proportion, this proportion being represented by the mean atomic weight of 35·46 given in the tables, which value accurately represents the *average* mass of the chlorine atom.

Symbolic notation is a necessity in the study of chemical changes, and for this object the atom of each element is represented by the first letter, or the first letter followed by another, of its English or Latin name. For example, S is the symbol for one atom of sulphur, Se for one atom of selenium and Sb for one atom of antimony (stibium).

The compound calcium carbonate (occurring as limestone, chalk and marble) consists of one atom of calcium, one of carbon and three of oxygen per molecule, its formula being written $CaCO_3$.

It has already been mentioned that calcium carbonate (marble), on being heated strongly, yields two substances, calcium oxide (quicklime) and carbon dioxide. This change is represented by an equation thus:

$$CaCO_3 = CaO + CO_2.$$

The equation therefore signifies that:

1. One molecule of calcium carbonate, consisting of one atom of calcium, one atom of carbon and three atoms of oxygen, yields one molecule of calcium oxide, consisting of one atom of calcium and one atom of oxygen, together with one molecule of carbon dioxide, consisting of one atom of carbon and two atoms of oxygen.

2. Applying the atomic weights, $Ca = 40$, $C = 12$, $O = 16$; $40 + 12 + 3 \times 16 = 100$ parts by weight of calcium carbonate yield $40 + 16 = 56$ parts by weight of calcium oxide and $12 + 2 \times 16 = 44$ parts by weight of carbon dioxide.

When quicklime is exposed to carbon dioxide the two substances readily unite, forming calcium carbonate once more; that is, this chemical change can be reversed, and this is indicated by writing the equation thus:

$$CaCO_3 \rightleftarrows CaO + CO_2,$$

the arrows signifying that the reaction can be made to proceed in either direction.

It should be noted that the number of molecules on each side of the equation is not necessarily the same, but the number of atoms of each kind must be. The *total mass* is unchanged.

As has been stated already, a chemical change which takes place spontaneously is accompanied by a liberation of energy, usually manifested and measured as heat. The above equations therefore are incomplete in so far as no account has been taken of the energy of the systems. Carbon dioxide unites with calcium oxide to form calcium carbonate with evolution of energy. It is called an *exothermic* change and the equation is more correctly written:

$$CaO + CO_2 = CaCO_3 + E.$$

The reverse reaction will of course require an equal quantity of energy to bring it about; it absorbs energy and is termed an *endothermic* change. The equation becomes:

$$CaCO_3 = CaO + CO_2 - E.$$

When carbon is burned in oxygen to form carbon dioxide, the full equation for the exothermic reaction becomes:

$$C + O_2 = CO_2 + E.$$

If one gramme-atom of carbon be taken (= 12 grm.), 32 grm. of oxygen will be required, of course, and the full equation becomes:

$$C + O_2 = CO_2 + 97,600 \text{ calories}.$$

These 97,600 calories therefore represent:

1. The heat of combustion of carbon (charcoal). One gramme-atom = 12 grm. being understood.

2. The heat of formation of carbon dioxide. That is, the heat evolved when one formula-weight in grammes of carbon dioxide ($12 + 2 \times 16 = 44$ grm.) is built up from its elements.

3. The heat of this reaction. That is, the reaction expressed by the substances and their weights in grammes represented by the equation.

The full statement then becomes: 12 grm. of carbon (charcoal) unite with 32 grm. of oxygen to form 44 grm. of carbon dioxide with the evolution of 97,600 calories. The reaction is exothermic.

Theory, Hypothesis, Law. A theory is an attempt to account for a particular set of phenomena which are related—a kind of explanation of their mechanism—and should fulfil two functions. In the first place it should enable the facts to be grasped as a whole, and should present a mental picture of the phenomena under consideration, and in the second it should enable deductions to be made which can then be put to the test of experiment. To ask whether a theory is true is beside the point: the question is rather, is it helpful? Often a theory has been propounded, such as the phlogiston theory for instance, which for a time has been useful, but later has led to conclusions contradicted by experience. A theory which once leads to unmistakably wrong conclusions is discarded, like a worn-out piece of machinery. Such a piece of machinery can sometimes be repaired and again be made to do good work. So can a theory, and Prout's theory, that all matter is composed of hydrogen atoms as the ultimate particles, is not a bad example. It was very attractive at first. Later, careful atomic weight determinations led to its being discredited and discarded, but recent work on the structure of the atom has reinstated the theory in an amended form (see Chap. XII).

A hypothesis is hardly distinguishable from a theory. It has been said that a theory is a conjecture which we hope may be true, and a hypothesis one which we hope may prove useful.

A theory or hypothesis which has stood the test of time and of crucial experiment, which has, in short, proved itself to be always in the right, is often raised to the dignity of a law. Sometimes this has been done prematurely, much to the discredit of the term, as in the case of the 'law' of artiads and perissads, or of odd and even valencies (Odling, 1864).

Avogadro's conjecture that equal volumes of all gases under the same conditions of temperature and pressure contain the same number of molecules, is really a theory or hypothesis. It is sometimes called Avogadro's law, though it is recognized that it is not exactly true. As a working hypothesis it has been invaluable to chemists. Modern chemistry, it is said, is founded on Avogadro's hypothesis.

It will be noticed that the ordinary use of the term in scientific literature implies two types, viz. *exact* laws and *approximate* laws. The former are confined to summaries of observed facts which have been verified and confirmed by each subsequent refinement of observation and experimental method over a long period. The latter are the more numerous, and the description is almost a contradiction in terms. The gas laws (p. 64) are approximate laws. They are very nearly true over small ranges of temperature and pressure, but fail over wider ranges. Even the law of constant proportions, on which all the quantitative work of the chemist depends, may prove to be approximate only. If the isotopes of chlorine should ever be separated there would be at least two kinds of sodium chloride, and it is not impossible that these should prove to be identical in specific properties; that their only observable difference may be that they do *not* contain the same elements united in the same proportions. Something closely approaching to this has already been observed in the case of the isotopes of lead (Soddy and Hyman, 1914).

Essential Features of Chemical Change. We have seen that chemical action is accompanied by changes of various kinds, which may be summarized as follows:

1. There is a disappearance of some specific properties and an appearance of new specific properties. Specific

properties may be defined as those properties which persist with the body, and whose sum makes up the concept of the body; that is, properties which cannot be changed. The properties which *can* be changed, such as size, shape, temperature, etc., are arbitrary or accidental properties. Specific properties which can be measured or given a numerical value are *constants*. Substances which have the same specific properties are chemically alike; that is, they are the same chemical individual.

2. The relative reacting masses are fixed and unalterable for the same change.

3. The change is accompanied by either an evolution or an absorption of energy.

4. There is sometimes a change of state. The *gases* hydrogen and oxygen unite to form a *liquid*, water. The *solid*, marble, on heating yields quicklime, *solid*, and a *gas*, carbon dioxide. If equal volumes of the *gases* ammonia and hydrogen chloride be allowed to mix, a *solid*, ammonium chloride, is formed and no gas remains. If a little water be added to chloral the two *liquids* unite to form a *solid*.

CHAPTER II

WATER, HYDROGEN, OXYGEN, OZONE, HYDROGEN PEROXIDE

WHEN viewed in layers of moderate thickness by transmitted light pure water is a colourless transparent liquid, but when seen through layers of great thickness it possesses a faint greenish blue colour. The colour of pure water has, however, little or no relation to that of certain lakes and seas, where the presence of foreign substances has sometimes a very marked effect.

Water is compressible in a very small degree, one million volumes being diminished by about fifty volumes when the pressure is increased from one to two atmospheres. Small as this compression is, its effect may be considerable. It has been calculated that the compression of water by the pressure of the atmosphere results in a lowering of the general sea-level by rather more than one hundred feet. If water were incompressible a large part of East Anglia, for instance, would be below the surface of the North Sea.

This compressibility can be shewn qualitatively in a very simple way. A litre flask, the neck of which has been drawn out to a fine capillary, is filled with coloured water, placed on a greased glass plate and covered with a bell jar fitted with a cork and rubber tube. On blowing suddenly through the tube a momentary depression of the water in the capillary can easily be seen.

The alteration of the specific volume of water with change of temperature is remarkable. The maximum density is attained at 4° C. Water at this temperature expands both on heating and cooling. The volume change of water with temperature variation is shewn in the following table.

Temp. °C.	Specific volume	Specific gravity	Temp. °C.	Specific volume	Specific gravity
0	1·000132	0·999868	7	1·000071	0·999929
1	1·000073	0·999927	8	1·000124	0·999876
2	1·000032	0·999968	9	1·000192	0·999808
3	1·000008	0·999992	10	1·000273	0·999727
4	1·000000	1·000000	20	1·001773	0·998230
5	1·000008	0·999992	50	1·01207	0·98807
6	1·000032	0·999968	100	1·04343	0·95838

It will be seen from the table that the volume changes between 0° C. and 10° C. are small, but they are of very great importance in the economy of nature. The bottom temperature of all deep seas, whether Arctic or tropical, is approximately 4° C., and this must have an enormous influence on the direction and strength of ocean currents. Were it not for this contraction with rise of temperature from 0° C. to 4° C., it is no exaggeration to say that the climate of the whole world would bear no sort of relationship to existing conditions.

The expansion of water on heating above 4° C. is small at first but increases with rise of temperature. If an ordinary litre flask be filled half-way up the neck with ice-cold water, on warming, the level of the water at first falls noticeably. This fall, however, is not entirely, or even mainly, due to contraction of the water, but to the more rapid expansion of the glass. As the temperature reaches the neighbourhood of 50° C. the water level rises rapidly. The water is now expanding more than the flask, and will probably overflow before the boiling point is reached.

Equilibrium between Liquid and Vapour. When freely exposed to the air, water (and ice) evaporate at a rate which increases with rise of temperature.

If an ordinary mercury barometer be prepared by filling a tube, closed at one end, with mercury and inverting it over a vessel of the same liquid, the level of the mercury in the tube being carefully noted, and a few drops of water admitted by means of a curved pipette to the top of the mercury column, the latter is seen to fall by an amount which depends on the temperature, and on the temperature only. If the water layer above the mercury is not very thick its weight can be neglected in what follows.

Fig. 1 shews two mercury barometers, one containing a little water above the mercury, which has sunk to B'. The pressure at B is equal to the pressure at B' since the levels are the same. The pressure at B' is due to the water vapour. The pressure at B is that of the mercury column AB. Therefore the pressure of water vapour (at the temperature of the experiment) is represented by AB millimetres of mercury.

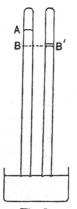

Fig. 1

If the barometer containing the water be heated, the mercury level will fall still further, always assuming that there is sufficient water present to maintain a layer of liquid water above the mercury, and at a particular temperature the level of the mercury in the tube will coincide with that in the dish. This temperature is the boiling point of water.

The Boiling Point of a Liquid is the Temperature at which the Pressure of its Vapour is equal to the External Pressure. The boiling point of a liquid therefore depends upon the external pressure. Water boils at 100° C. only when the external pressure equals 760 mm. of mercury.

The following table gives the values of the vapour pressure of water at various temperatures (Regnault).

Pressure of Aqueous Vapour in mm. of Mercury

Temp. °C.	mm.	Temp. °C.	mm.	Temp. °C.	mm.	Temp. °C.	Atmos.
−10	2·08	16	13·54	90	525·39	100	1·0
− 9	2·26	17	14·42	95	633·69	110	1·4
− 8	2·46	18	15·36	99	733·21	120	1·96
− 7	2·67	19	16·35	99·1	735·85	130	2·67
− 6	2·89	20	17·39	99·2	738·50	140	3·57
− 5	3·13	21	18·50	99·3	741·16	150	4·7
− 4	3·39	22	19·66	99·4	743·83	160	6·1
− 3	3·66	23	20 89	99·5	746·50	170	7·8
− 2	3·96	24	22·18	99·6	749·18	180	9·9
− 1	4·27	25	23·55	99·7	751·87	190	12 4
0	4·60	26	24·99	99·8	754·57	200	15·4
1	4·94	27	26·51	99·9	757·28	210	18·8
2	5·30	28	28·10	100	760·00	220	22·9
3	5·69	29	29·78	100·1	762·73	230	27·5
4	6·10	30	31·55	100·2	765·46		
5	6·53	35	41·83	100·3	768·20		
6	7·00	40	54·91	100·4	771·95		
7	7·49	45	71·39	100·5	773·71		
8	8·02	50	91·98	100·6	776·48		
9	8·57	55	117·48	100·7	779·26		
10	9·17	60	148·79	100·8	782·04		
11	9·79	65	186·94	100·9	784·83		
12	10 46	70	233·08	101	787·59		
13	11·16	75	288·50	105	906·41		
14	11·91	80	354·62	110	1075·37		
15	12·70	85	433·00				

A difference of 1° C. in the boiling point of water (under ordinary conditions of atmospheric pressure) corresponds to a difference of approximately 27 mm. in the height of the barometer, therefore the boiling point of water at sea-level may easily fluctuate between 99·5° C. and 100·5° C. with variations in barometric height. Water very seldom boils at exactly 100° C.

This variation in the temperature of boiling water has been used to determine the heights of mountains; or rather, the barometric pressure, from which the height can be deduced. On the summit of Mont Blanc water boils at a temperature of approximately 80° C.

As will be seen from the table, the vapour pressure of water increases very rapidly with rise of temperature above 100° C. When a high-pressure steam boiler shews a head of steam of 150 lb. to the square inch (10 atmospheres) the temperature of the water is slightly above 180° C. In fact the pressure inside a steam boiler is more accurately indicated by a thermometer (protected by a steel tube) than by a pressure gauge working against a steel spring.

Critical Phenomena. Let us imagine a thick-walled glass tube of small bore, capable of withstanding high pressures, to be partly filled with water and sealed up. If the tube be heated the vapour pressure inside will increase; that is, more liquid will pass into the gaseous state, while the *volume* of the vapour will not greatly change. The *density* of the vapour becomes greater. If the temperature be continuously raised, a point will at length be reached when the density of the vapour will equal that of the liquid. At this stage the contents of the tube are homogeneous—equal volumes at all parts of the tube contain the same number of molecules—and the surface between vapour and liquid disappears. The contents of the tube are entirely gaseous. This marks the critical temperature; the temperature above which a gas cannot be liquefied by any pressure, however great.

The *critical temperature* then is the temperature above which the substance is gaseous at all pressures. This is sometimes called the 'absolute boiling point' (Mendeléeff. 1861), but the expression is misleading (see *absolute zero*,

p. 65). The *critical pressure* is the vapour pressure exerted
by a liquid at its critical temperature.

	Critical temperature	Critical pressure
Water	365° C.	200·5 atmospheres
Carbon dioxide	30·9° C.	70 ,,
Oxygen	−118·8° C.	50 ,,
Hydrogen	−241° C.	20 ,,

Superheated Water. Under certain conditions water can
be heated above its normal boiling point, but the liquid
is then unstable, and is liable to pass into the gaseous condition suddenly and with violence. Distilled water heated
in a clean glass vessel often becomes slightly superheated,
and then boils with succussion or 'bumps.' To avoid this
it is usual to place a few fragments of some insoluble
material in the vessel, such as glass beads, porous pipe-
clay, or scraps of platinum. The irregular surfaces facilitate
the formation of bubbles of steam and the water boils
smoothly.

Superheating may be readily shewn by shaking a drop
of water into an emulsion with a few cubic centimetres of
olive oil. The emulsion can be heated to a temperature far
above 100° C., when a crackling noise will result as the
tiny droplets of superheated water suddenly pass into
steam—the noise which accompanies the operation of
frying in fat.

Desiccation. In experimental work it is frequently
necessary to deal with substances which must be dry.
Sparingly soluble gases are usually collected over water,
but are then saturated with water vapour. Many solids
and liquids are more or less hygroscopic, and water must
be removed without altering the substance in any other
way. The materials which are most commonly used as
desiccating agents are concentrated sulphuric acid, an-

hydrous calcium chloride, calcium oxide (quicklime), and phosphorus pentoxide, all of which rapidly absorb water or water vapour. In making a selection of one or other of these drying agents it is essential that the substance which is to be dried shall not react chemically with it. For instance, hydrogen may be dried by allowing the gas to bubble slowly through concentrated sulphuric acid, or by passing it through tubes filled with anhydrous calcium chloride, but ammonia cannot be dried in either of these ways because it combines with both sulphuric acid and calcium chloride. Water vapour may, however, be removed from ammonia by passing the gas through towers filled with lumps of quicklime.

Solids which have to be weighed after heating must not be allowed to cool while exposed to the atmosphere, because in cooling they are liable to absorb atmospheric moisture; the cooling must take place in closed vessels, known as desiccators, containing one of the substances above mentioned.

Liquids are dried by adding some insoluble drying substance, the liquid of course being kept in a closed vessel. Anhydrous (or absolute) alcohol may be prepared from rectified spirit, which contains some water, by quicklime. The spirit is poured upon about half its weight of quicklime in a stoppered flask and allowed to stand for some days. The alcohol may then be poured, or filtered, or distilled off. Calcium chloride both combines with, and dissolves in, alcohol. It is therefore unsuited for drying this liquid.

Equilibrium between Liquid and Solid Water. At temperatures below 0° C. ice is the stable modification of water. At 0° C., under atmospheric pressure, the solid and liquid co-exist in equilibrium. It is possible to cool water a few degrees below 0° C. without the formation of ice, but the liquid under such conditions is not stable, but *meta-stable*. If a fragment of ice be introduced into water thus supercooled more ice separates at once, and at the same time the temperature rises to 0° C. The range of temperature below 0° C. through which water can remain liquid in the absence of ice is however limited, and is dependent upon the experimental conditions. The greater the degree of supercooling the greater is the tendency for water to solidify, or in other words, the less stable does the liquid become. If water be cooled below a certain temperature, formation of ice takes place without the introduction of a particle of the solid; the metastable range of temperature is passed.

When water passes from the liquid to the solid state the change is accompanied by a considerable increase (about one-eleventh) in volume. This increase in volume is remarkable. Most liquids contract on solidifying; those which expand being very exceptional. In no ordinary case is the expansion on solidification at all comparable with that which accompanies the conversion of water into ice. When water is cooled below 0° C. in a closed vessel capable of withstanding great pressure, in such a way that this expansion cannot take place, ice is not formed; the water remains liquid. If the pressure be removed, ice is instantly formed.

Various aspects of this phenomenon are of importance in nature. Everyone is familiar with the bursting of water pipes in cold weather. The weathering of rocks is partly due to the filling of the interstices with water, and the disruptive pressure caused by the freezing of this water.

The phenomenon of the melting of ice is in some respects simpler than that of the freezing of water. Although it is possible to supercool a liquid below its freezing point, it is not possible to heat ice above 0° C. without melting it,

It was shewn theoretically by J. Thomson in 1849, that in the case of a substance such as water, in which the specific volume is greater in the solid than in the liquid state; that is, a liquid which expands on solidification; increase of pressure should lower the melting point. In the case of water the lowering should be approximately 0·0075° C. for every additional atmosphere. This prediction was verified experimentally in the following year by W. Thomson (Lord Kelvin), who found that under a pressure of 8·1 atmospheres the melting point was − 0·059° C., and under a pressure of 16·8 atmospheres the melting point was − 0·12° C. The pressure necessary to produce a lowering in the freezing point of 1° C. is 134 atmospheres.

The influence of pressure on the melting point of ice can be shewn in a qualitative manner by placing a fine, strong wire round a block of ice fixed in a horizontal position, the two ends of the wire holding a scale pan loaded with heavy weights. The wire gradually cuts its way through the block of ice, but the block is not severed; the water freezes again because the pressure due to the wire, which causes the ice immediately under it to melt, is released, and the freezing point of the water is once more raised. This phenomenon is known as regelation, and was discovered by Faraday in 1850. Tyndall considered that regelation is responsible for glacier motion, but there is no doubt that the viscosity of ice must also be taken into account. The 'slipperiness' of ice is also due in some measure to this phenomenon.

Ice has a definite vapour pressure, the value of which, like that of liquid water, depends on the temperature. The production of hoar frost is said to be due to the direct passage from the vapour phase to the solid without the intermediate formation of liquid. The vapour pressure of ice is always lower than that of supercooled water at the same temperature. This is a general phenomenon; the less stable form of any substance has always the greater vapour pressure.

Equilibrium between the Three States of Aggregation of Water. We have seen that, under atmospheric pressure, ice melts at 0° C. At this temperature the pressure of water vapour, and therefore of ice with which it is in equilibrium, is 4·6 mm. of mercury. If the system be considered as under the pressure of water vapour only; that is, without the atmospheric pressure; it can be shewn both theoretically and experimentally that at a temperature of + 0·0075° C. and at a water vapour pressure of 4·6 mm. of mercury; ice, liquid water, and water vapour will coexist in stable equilibrium. This temperature and pressure constitute what is termed the *triple point* of water, and only at the triple point can the three states of aggregation, or *phases* as they are usually called, remain together in equilibrium.

The relations between temperature and pressure and the various phases of water are most conveniently studied with the aid of an equilibrium diagram. Temperatures are plotted as abscissae with corresponding pressures as ordinates. The line OA is the vapour pressure curve of liquid water, OB that of ice, and OA' that of supercooled water; the line OA' is above OB because the vapour pressure of

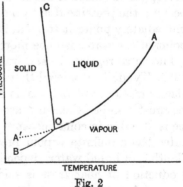

Fig. 2

supercooled water is always greater than that of ice at the same temperature, as has already been shewn. The line OC represents the influence of pressure on the

* This section may be omitted on a first reading.

melting point of ice. It is slightly inclined towards the pressure axis because the effect of increase of pressure is to lower the melting point. The point O is the triple point. The three lines OA, OC and OB represent respectively, equilibrium between liquid and vapour, liquid and solid, and solid and vapour. In every case, if either the temperature or pressure be fixed, the other condition assumes a definite value. This is usually expressed by saying that the system under those conditions has one degree of freedom. The three regions enclosed within these lines represent areas in which both the temperature and the pressure may be varied, but only one phase can be present. Such regions are said to have *two* degrees of freedom.

The conditions of equilibrium between ice, water, and steam are a special case of a general rule which deals with equilibrium in heterogeneous systems, developed theoretically by Willard Gibbs (1874–1878). This *Phase Rule* may be expressed by the equation

$$P + F = C + 2,$$

where P denotes the number of phases, F the number of degrees of freedom, and C the number of components. In the system under consideration, there is only one component, namely, the substance water: the various phases of the substance water have all the same chemical composition. At the triple point there are no degrees of freedom, and three phases are present; such a system is termed non-variant. Along any of the lines OA, OB and OC there is one degree of freedom, and two phases are present; a system of this kind being termed univariant. In the regions between these lines, only a single phase is present; the system has two degrees of freedom and is called a bivariant one.

It is very important for beginners to understand precisely how much information the phase rule gives. The rule tells us only *how many* phases can coexist under certain specified conditions: it does not tell us what the phases are, nor does it tell us which particular phase will appear or vanish if the conditions of equilibrium are altered. The question as to what the phases are, in any given case, is a matter for experiment; and the question of the appearance or disappearance of any particular phase is governed by a general principle known as Le Chatelier's theorem (1884). This principle may be stated in the following terms. *If a system in equilibrium is subjected to a constraint by which the equilibrium is shifted, a reaction takes place which opposes the constraint; that is, one by which the effect of the constraint is partially annulled.*

If we apply Le Chatelier's theorem to the equilibrium between ice and liquid water, the effect of raising the temperature will be to cause a reaction to take place with an absorption of heat. The melting of ice is an endothermic process; application of heat will result therefore in some of the ice being transformed into water. The temperature will not rise however until the whole of the ice has been melted, equilibrium being independent of the *amounts* of the phases. A definite quantity of heat is required to melt a given quantity of ice, 80 calories being absorbed for every gramme of ice which is transformed into liquid water at 0° C. This is sometimes expressed by saying that the latent heat of fusion of ice is 80 gramme-calories. The quantity of heat which is required to raise the temperature of one gramme of water through a range of temperature of 1° C. is termed one calorie, but in exact work it is necessary to specify the particular temperature at which the calorie is defined, as the specific heat of water, like that of other substances, varies with the temperature. The absorption of heat in the process of melting a solid is a general phenomenon and was discovered by Black in 1762.

As in the case of the transformation of a solid into a liquid, so in the case of the transformation of a liquid into a vapour at the same temperature, heat has to be supplied. The quantity of heat which is required to convert one gramme of water at 100° C. into steam at the same temperature, or, as it is usually termed, the latent heat of vaporization of water, is 536 calories.

Aqueous Solutions. Water has remarkable solvent properties: no other liquid is comparable with it as regards its solvent action. For this reason water is never met with in nature in a state of purity. Even rain-water, which is the purest variety of natural water, contains dissolved gases. Other natural waters contain varying quantities of dissolved solids.

If a soluble salt such as potassium nitrate be added to water, the solid gradually disappears. If the liquid be stirred a homogeneous solution is obtained, and such a solution is permanent; that is, there is no tendency towards segregation into layers of varying concentration. If successive quantities of the salt be added, a point is ultimately reached when the liquid is incapable of dissolving any more; however long it be stirred, the excess of salt simply remains undissolved. When this stage has been reached the solution is saturated: equilibrium between the solution and the solute has been established. If the solution be heated, more of the undissolved solid passes into solution and on cooling it crystallizes out.

The *solubility* of a substance in water is an expression of the mass of that substance which is dissolved by a definite quantity of water under specified conditions. Excess of the solid must be present, as otherwise there is no assurance that the solution is saturated. The *value* of the solubility can be expressed in different ways; the most important modes of expression being either to say that

100 parts by weight of water can dissolve so much of the substance, or that the solution contains so many grammes of the substance per litre of solution. In all cases it is essential that the temperature be stated, since solubility is a function of the temperature.

Under particular conditions it is possible to prepare *supersaturated* solutions; that is, solutions which contain more of the solute than corresponds with the condition of saturation. The phenomenon may be shewn thus: If 100 grm. of washing soda be dissolved in 40 grm. of water with heating, the hot solution filtered into a clean glass flask and the neck then plugged with cotton wool to keep out dust, on slowly cooling to laboratory temperature a clear homogeneous liquid is obtained. This solution is metastable, as may be shewn by the introduction of a minute crystal of washing soda. At once a considerable separation of crystals takes place, with a rise in temperature. Highly supersaturated solutions may sometimes be caused to crystallize by violently shaking, or by scratching the interior of the vessel.

A supersaturated solution resembles a supercooled liquid in many respects. Both are metastable; that is to say, stable in the absence of the solid phase.

The solubility of most solids increases with rise of temperature, though some, notably certain calcium salts, are less soluble in hot water than in cold. Frequent use is made of solubility curves in experimental work. These curves are obtained by determining analytically the concentration of the salt in saturated solutions at various temperatures and expressing the results graphically by plotting the concentrations against the corresponding temperatures. A solubility curve is usually drawn for 100 grm. of the solvent, the ordinates representing the mass of solute, while the abscissae represent the temperatures.

In Fig. 3 the point X indicates that for this particular solute a saturated solution at the temperature OA contains OB grm. of the solute dissolved in 100 grm. of the solvent. All points on the line XY represent the condition of equilibrium between the solution and the solute. The point Q is in the metastable or unstable region, and if it could be attained experimentally (by dissolving QT grm. of solute in 100 grm. of water at a temperature above T' and

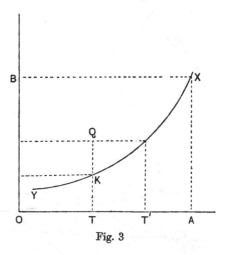

Fig. 3

cautiously cooling down to T) would represent a supersaturated solution. Such a solution, if caused to crystallize, would deposit QK grm. of solute and leave a saturated solution containing KT grm. of solute dissolved in 100 grm. of water at the temperature $OT°$ C.

Fig 4 shews the solubility curves of some well-known salts.

The solubility curves of some solids are continuous; in other cases, that of sodium sulphate for example, the curve has one or more abrupt discontinuities. The meaning of a discontinuity in a solubility curve is the appearance of a

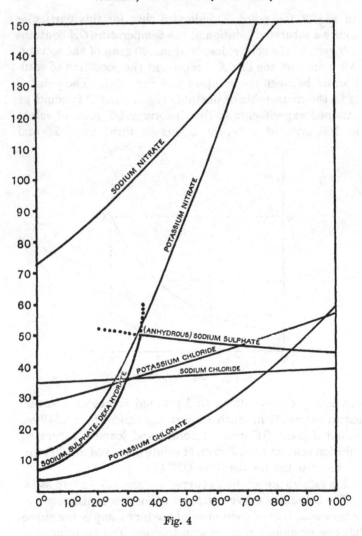

Fig. 4

new solid phase in contact with the saturated solution. Two types of salts may be distinguished, namely, those which crystallize from their solutions as anhydrous salts, and those which crystallize as hydrated salts; that is, salts with one or more molecules of water in chemical combination. Potassium nitrate belongs to the former class and sodium sulphate to the latter. At temperatures below 33° C. sodium sulphate crystallizes with ten molecules of water to every molecule of sodium sulphate. The solubility curve of such a salt consists essentially of two parts; one part representing the equilibrium between water and the decahydrate up to 33° C., the other part the solubility of the anhydrous salt from 33° C. upwards. A saturated solution of sodium sulphate at 33° C. would deposit crystals on either heating or cooling, the maximum solubility being at this temperature. On heating such a solution the anhydrous salt, Na_2SO_4, would separate; on cooling the decahydrate, $Na_2SO_4.10H_2O$, crystallizes out. The temperature 33° C. is a transition point, since both solid phases, viz. anhydrous sodium sulphate and the decahydrate, are in equilibrium with the saturated solution. It is possible to study the solubility of the anhydrous salt at temperatures slightly below this, and also the solubility of the decahydrate over a small range above 33° C. as shewn by the dotted lines in Fig. 4, but in both cases the determinations are made in metastable regions. In the diagram the ordinates represent the masses of both solutes, calculated as anhydrous salts, dissolved by 100 grm. of water at the corresponding temperatures.

A study of solubility curves is of importance in the process of purifying solids by recrystallization. Suppose the solubility of a salt is X grm. per 100 grm. of water at $T°$ C., and x grm. at some lower temperature $t°$ C. For simplicity it will be assumed that the curve is continuous, that is, that no hydrates are formed. If a saturated solution be prepared at T° C. and the solution be cooled to t° C. $X - x$ grm. of salt will crystallize out. Suppose now that the salt contains a small quantity of some soluble impurity, insufficient to saturate the solution at t° C. On making a saturated solution at T° C. of the impure salt and cooling to t° C. the crystals deposited will not contain the impurity, which will remain in solution. It is seldom that the problem of purification is quite so simple as this, usually more than one recrystallization being necessary.

Some substances dissolve in water with absorption of heat, others with evolution of heat. If a test tube be half filled with ammonium nitrate and the latter just covered with water the fall in temperature is considerable, and may be sufficient to freeze a film of water on the outside of the tube. On the other hand, when sodium hydroxide is dissolved in water there is a considerable evolution of heat. In general, hydrated salts dissolve with absorption of heat; anhydrous salts which are capable of forming hydrates usually dissolve with evolution of heat. The problem is, however, somewhat complicated and may be studied in works on physical chemistry.

Solubility of Liquids. Some liquids, such as alcohol and acetone, are completely miscible with water; i.e. they are both infinitely soluble in water. Others, such as mercury and petroleum, are practically non-miscible with water. There are also many liquids, such as ether and chloroform, which mix with water to a limited extent. If, for example, phenol be gradually added to water it goes into solution, but after a certain quantity has been added two layers are formed, the upper layer consisting of water saturated with phenol, and the lower of phenol saturated with water. As the temperature is gradually raised the mutual solubility increases until $68 \cdot 5^\circ$ C. is reached, when the two layers become one; that is, at and above this temperature, known as the critical solution temperature, phenol and water are infinitely soluble in each other. On cooling, the phenomena are reversed.

The relations between phenol and water may be illustrated by the accompanying diagram. At the temperature OA the solubility of phenol in water has two values:

(a) AP grm. of phenol dissolve in $A'P$ grm. of water, and (b) AP' grm. of phenol dissolve in $A'P'$ grm. of water.

This will be better understood by regarding (a) as the solubility of phenol in water and (b) as the solubility of water in phenol, as above expressed.

OB represents the critical solution temperature, 68·5° C.

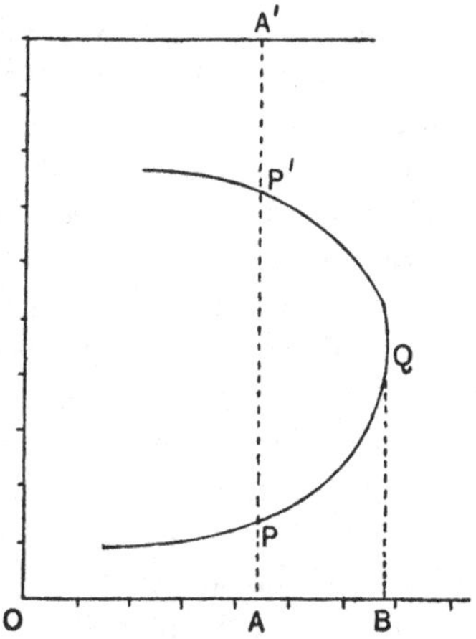

Fig. 5

Most liquids of limited miscibility give solubility curves of this type. In some few cases the diagram is reversed—the mutual solubility diminishes with rise of temperature. Trimethylamine is a good example of this. If equal volumes of trimethylamine and water be shaken together at ordinary laboratory temperature a homogeneous liquid is produced. On warming, even at the temperature of the hand—or better, by placing the tube in a vessel of warm water—two layers quickly separate. On cooling, the liquid once more becomes clear and homogeneous.

Solubility of Gases. Gases differ greatly in their solubility. Hydrogen and nitrogen for instance dissolve very sparingly, while ammonia and hydrogen chloride are extremely soluble in water. The solubility of all gases depends on temperature and pressure; in the case of the sparingly soluble gases the quantitative relations are governed by a law stated by Henry (1803). This law may be expressed thus: *The mass of a gas which is absorbed by a liquid is directly proportional to the pressure of the gas so long as the temperature is constant*. Since, according to Boyle's law, the volume of a gas is inversely proportional to the pressure, Henry's law may also be expressed as follows: *A given volume of a liquid will absorb the same volume of a gas at any pressure of that gas so long as the temperature is constant*.

The highly soluble gases do not follow any simple law.

The solubility of all gases (with the somewhat doubtful exception of helium over a short range) diminishes with rise of temperature. Bunsen defined the coefficient of solubility of a gas as the volume of the gas, calculated at 0° C. and 760 mm. pressure, which is absorbed by unit volume of the solvent at a pressure of 760 mm. of the gas at the stated temperature. This coefficient has however been gradually replaced by a simpler one due to Ostwald, who defined the solubility of a gas as v/V, where v denotes the volume of gas absorbed by V volumes of solvent at the stated temperature, a statement of the pressure being unnecessary, as has been seen in the second form of expressing Henry's law.

Dalton (1805) observed that the pressure of a gaseous mixture is equal to the sum of the partial pressures of each of the gases present. Thus, in the case of the two principal constituents of the atmosphere, oxygen and nitrogen, which are present in the approximate proportions by volume of one part of the former to four of the latter, a total atmospheric pressure of 760 mm. would be represented approximately by 152 mm. due to the oxygen and

608 mm. due to the nitrogen. Since the coefficient of solubility of oxygen in water at 0° C. is 0·04 and that of nitrogen 0·02, ice-cold water saturated with air should contain in each cubic centimetre:

$$\text{oxygen, } \frac{0\cdot04 \times 152}{760} = 0\cdot008 \text{ c.c.};$$

$$\text{nitrogen, } \frac{0\cdot02 \times 608}{760} = 0\cdot0158 \text{ c.c.};$$

the gases being measured at standard temperature and pressure.

Again, the coefficient of solubility of carbon dioxide at 15° C. is approximately unity; i.e. water at ordinary temperatures, under a pressure of one atmosphere of carbon dioxide, dissolves about its own volume of the gas. Since the proportion of carbon dioxide in the atmosphere is only three parts in ten thousand, the pressure of it is about 0·25 mm. of mercury, therefore one cubic centimetre of pure water at 15° C. only dissolves $\frac{0\cdot25}{760} \times 1$ c.c. of carbon dioxide from the atmosphere, i.e. 0·0003 c.c.

In certain districts there are found springs the water of which may contain two or three times its own volume of carbon dioxide. This water has dissolved the gas under conditions where the pressure of carbon dioxide is high, but on reaching the surface, where the pressure of this gas is very small, equilibrium is restored by a rapid escape with effervescence. Many so-called mineral waters and soda waters are merely water saturated with carbon dioxide under pressure.

Natural waters differ greatly in the amounts of the substances they contain in solution. Mountain streams which flow over igneous rocks usually contain but little solid matter. The Scottish lochs are remarkably free from dissolved solids. On the other hand, lakes which have no outlet, such as the Dead Sea, Lake Elton, and the Great Salt Lake, where the only loss of water is by evaporation, usually contain very considerable quantities of solids. The deeper waters of the Dead Sea, for instance, are saturated with common salt. Oceanic waters contain on an average about 3 per cent. of solid matter, mostly common salt; the Mediterranean considerably more than this and the Baltic less. 'Mineral' waters are those which contain much solid (or gas) dissolved in them: *hepatic* springs, such as those of Harrogate, contain hydrogen sulphide; *chalybeate* springs, such as those of Bath and Tunbridge Wells, are charged with iron in the form of the bicarbonate, while *petrifying* springs contain much calcium bicarbonate.

Hard Waters are those which do not easily form a lather with soap, in consequence of the presence of certain substances in solution, the commonest being the salts of calcium and magnesium. Soap is a mixture of the sodium or potassium salts of certain fatty acids, which react with the dissolved salts to form sparingly soluble calcium or magnesium salts of those acids. These are precipitated as a curd, and the formation of a lather cannot take place until the whole of the calcium has been removed from solution. The reaction between calcium sulphate in aqueous solution and sodium stearate (a typical constituent of soap) may be represented thus:

calcium sulphate + sodium stearate = calcium stearate + sodium sulphate.
  (soluble)   (soluble)   (insoluble)   (soluble)

Two kinds of hardness may be distinguished, *temporary* and *permanent*. Temporary hardness is due to the presence of certain bicarbonates, usually those of calcium, magnesium or iron, in solution. Calcium carbonate, $CaCO_3$, is

practically insoluble in water, but in presence of carbon
dioxide it goes into solution as the bicarbonate, thus:

$$CaCO_3 + H_2O + CO_2 \rightleftharpoons Ca(HCO_3)_2.$$

This reaction is reversible as shewn in the equation,
calcium bicarbonate being stable only in presence of carbon
dioxide. All waters with hardness of this kind become
cloudy on prolonged exposure to the air, the surface of the
containing vessel becoming covered in time with a film of
calcium carbonate. If such a solution be boiled the bicarbonate is quickly decomposed, the carbon dioxide
escaping into the air and the carbonate being deposited.
Temporary hardness is defined as hardness which can be
removed by boiling. The 'furring' of kettles and boilers,
which is so well known in districts where the water supply
comes from chalky soil, is due to this decomposition.

Another interesting aspect of this reaction is to be seen
in petrifying springs where the water, charged with calcium
bicarbonate under a high pressure of carbon dioxide, falls
in drops from the roof of a cave. The hanging drop loses
carbon dioxide and leaves a tiny pellicle of calcium carbonate adhering to the roof. These particles accumulating
through long ages form an inverted pinnacle, or *stalactite*,
and on the ground where the drop has fallen another pinnacle, or *stalagmite*, grows upwards and the two may
ultimately meet to form a column. These formations are
to be seen in their greatest size in the Mammoth Caves of
Kentucky, but are also met with in the caves of Cheddar,
Han (Belgium), and S. Beatenberg (Switzerland), and indeed
in nearly all limestone districts. Objects placed in such
waters form nuclei on which the separation of chalk takes
place easily. They become, in fact, centres of crystallization,
and the petrifying consists in their becoming covered with
a stony deposit more or less thick.

Methods other than boiling are sometimes used for removing the hardness due to calcium bicarbonate. On the addition of slaked lime (calcium hydroxide, $Ca(OH)_2$) in the form of milk of lime, the following reaction takes place:

$$Ca(HCO_3)_2 + Ca(OH)_2 = 2CaCO_3 + 2H_2O,$$

the calcium being precipitated in the form of insoluble carbonate which settles out in the reservoir. This is known as Clark's process and is still in general use.

In practice the whole of the temporary hardness is never removed from the water of town supplies. Water free from hardness, although very acceptable for washing purposes and for steam raising, is unpalatable, and may sometimes cause dysentery. Pure water too has a quite considerable action on lead, and on its passage through leaden pipes may take into solution sufficient lead in the form of the hydroxide, $Pb(OH)_2$, to cause serious lead poisoning. Water containing calcium bicarbonate quickly covers the surface of the pipe with a coat of insoluble carbonate which prevents further action on the metal.

A method of partially softening temporarily hard water sometimes employed by small steam users is the following:

A quantity of ammonium chloride is dissolved in the feed water of the boiler. In the boiler itself this reacts with the bicarbonate thus:

$$2NH_4Cl + Ca(HCO_3)_2 = CaCl_2 + 2NH_3 + 2CO_2 + 2H_2O,$$

the ammonia and carbon dioxide passing off with the steam, and the very soluble calcium chloride being left in the water. By occasionally blowing out and replacing the water the formation of 'fur' is diminished but not altogether prevented. This method has serious disadvantages however, since the presence of even small quantities of ammonium chloride in boiler water leads to rapid corrosion and pitting of the plates.

The action of iron on water containing bicarbonate is remarkable. A stream of such water passed through a tube filled with iron turnings loses temporary hardness. It depends of course on the length of time that the water is in contact with the iron, but the whole of the temporary hardness can be removed in this way. Incidentally, the water is still further purified by the removal of bacteria. In an extreme experiment a specimen of sewage water was passed slowly through a long tube filled with rough steel turnings, when the collected water was found to contain fewer bacteria than ordinary drinking water.

Permanent hardness is defined as hardness which is unaffected by boiling, and is due to the presence of salts which are stable in a boiling solution, the commonest being the sulphates of calcium and magnesium. The softening of such water has therefore to be effected by some reaction for removing the calcium or magnesium by precipitating it as an insoluble salt. One method of so doing is to add sodium carbonate to the water, when calcium carbonate is precipitated.

$$CaSO_4 + Na_2CO_3 = CaCO_3 + Na_2SO_4,$$

the sodium carbonate being added in slight excess if all the calcium sulphate is to be removed.

Sodium carbonate can remove the hardness due to calcium bicarbonate as well as to the sulphate, the reaction involving the formation of calcium carbonate and sodium bicarbonate:

$$Na_2CO_3 + Ca(HCO_3)_2 = CaCO_3 + 2NaHCO_3.$$

Sodium carbonate is the active principle in so-called bath salts.

In recent years much progress has been made in methods for softening water depending upon decomposition of the dissolved calcium or magnesium salts with sodium silicates.

The material known as *permutite* is an artificial silicate of sodium and aluminium, a typical analysis of which is:

Silica (SiO_2)	46·0
Alumina (Al_2O_3)	22·0
Sodium (calculated as oxide, Na_2O)	13·6
Water	18·4
	100·0

Denoting for brevity the active constituent of this material as $Na_2(SiO_3)_x$, the changes which occur when calcium sulphate or bicarbonate reacts with it may be thus represented:

$$Na_2(SiO_3)_x + CaSO_4 = Na_2SO_4 + Ca(SiO_3)_x,$$

and

$$Na_2(SiO_3)_x + Ca(HCO_3)_2 = Ca(SiO_3)_x + 2NaHCO_3.$$

The permutite is contained in filters which are directly attached to the water supply, and when the sodium silicate is exhausted it can be very simply and easily regenerated by pouring brine through the apparatus:

$$2NaCl + Ca(SiO_3)_x = CaCl_2 + Na_2(SiO_3)_x,$$

the active constituent of the permutite being thereby renewed. The very soluble calcium chloride is simply washed out of the apparatus.

It will have been noted that some of the methods of softening described are inapplicable to potable waters.

The reverse process, the hardening of water, is sometimes practised, since as has been stated above, very pure waters are neither palatable nor safe. This may be done by allowing the water to flow over artificial chalk beds; the treatment applied to the water of certain districts where the hardness is seriously deficient.

The quality of the beer brewed at Burton-on-Trent is said to be due in part to the presence of calcium sulphate in the water used, and in other districts brewers harden the water supply with this salt with the object of improving their product.

Pure Water. Owing to its great solvent power the preparation of pure water is a matter of no small difficulty. It is well known that distilled water must be used for many operations in analytical chemistry, since the dissolved impurities in many natural waters would constitute a serious source of error. Ordinary distilled water though practically free from dissolved solids, still contains dissolved gases, these depending partly on the source of the water and also to some extent upon the manner in which the distillation is conducted. Ammonia is a particularly difficult gas to eliminate on account of its high solubility. Carbon dioxide, though very much less soluble, is also a source of trouble, since it is never absent from the atmosphere. This gas is derived chiefly from the decomposition of calcium bicarbonate which takes place when a water possessing temporary hardness is boiled, the gas being carried over with the steam and consequently to some extent redissolved in the distillate.

In experiments upon the electrical conductivity of aqueous solutions the use of water of a high degree of purity is of course a necessity, and simple distillation is not sufficient purification. Organic matter may be removed with potassium permanganate, the liquid being distilled with or without the addition of dilute sulphuric acid. The resulting distillate is next made alkaline with baryta water which retains carbon dioxide as insoluble barium carbonate, and the alkaline liquid again distilled. If the water contains ammonia, that impurity will not be eliminated unless the distillate from the treatment with baryta water is redistilled from an acid solution (potassium bisulphate). In all experiments where water of a high degree of purity is required, it is important to avoid the use of glass as much as possible on account of the solvent action of water on the alkali silicate in this material. Tin or silver condensers free from soldered joints are generally used, but in cases where the use of glass cannot be avoided, the vessels should be previously steamed for a long time to remove the soluble silicates as far as possible.

The criterion of purity universally employed in testing water is that of its electrical conductivity. Careful experiments have shewn that the conductivity of water steadily diminishes with decrease of impurity. Probably the purest water ever prepared was that obtained by Kohlrausch and Heydweiller (1894). These investigators distilled water *in vacuo*. They started by freezing ordinary distilled water and pouring off the unfrozen portion, which of course was less pure than the ice. This method of purification appears to have been first tried by Nernst (1891) but has been very little used since, although it has been frequently employed for the purification of other substances. The water obtained by melting the purified ice was then distilled *in vacuo*, the conductivity being determined directly in a vessel sealed on to the apparatus. The value of the conductivity of the purest water obtained was approximately 0.04×10^{-6} in reciprocal ohms at $18°$ C. This value is almost certainly a specific property of water, and cannot be ascribed to residual impurities. On exposure to the air for a very short time, water so purified rapidly absorbed gases and its conductivity rose in consequence. For ordinary physicochemical experiments there is no necessity whatever to work with water of the degree of purity realized by Kohlrausch and Heydweiller, and indeed the experimental difficulties of doing so would be very great. Water of a high degree of purity, free from both ammonia and carbon dioxide, may be obtained in one operation by distilling a dilute solution of Nessler's reagent. The latter is a solution of potassium mercuric iodide containing excess of potash and retains both these gases.

Electrolysis of Aqueous Solutions. Although pure water is almost a non-conductor of electricity, if it contains small quantities of certain classes of compounds in solution it becomes a conductor. If, for example, two platinum strips connected respectively to the two poles of a suitable battery be immersed in a vessel of pure water, no measurable current passes, but on the addition of a few drops of dilute sulphuric acid a current at once flows, and gases are evolved on the platinum strips. When the current is interrupted the evolution of gas ceases.

A liquid such as dilute sulphuric acid, which is capable of conducting electricity with simultaneous occurrence of chemical change, is termed an *electrolyte,* or conductor of the second class, to distinguish it from substances which conduct without chemical change, such as metals, which are conductors of the first class.

If an apparatus be arranged to collect the gases evolved, it will be observed that the volume which is obtained from one of the platinum strips is approximately double that from the other. Such an apparatus is known as a voltameter, and the metal strips as the electrodes. The electrode which is connected with the positive pole of the battery or other source of current is called the *anode,* the other the *cathode.* The volume of gas liberated at the cathode in this experiment is double that obtained from the anode (Fig. 6). Examination of the two gases shews that they possess very different properties. The cathode gas is hydrogen, the anode gas oxygen. Both these gases are prepared on a commercial scale by methods essentially similar to the one described above.

If the hydrogen and oxygen obtained by the electrolysis

of dilute sulphuric acid be mixed and dried, and then introduced into a strong glass vessel provided with two platinum wires sealed in, on passing an electric spark between the wires the two gases combine with explosive violence and a film of moisture makes its appearance on the walls of the vessel.

The experimental proof of the volume composition of water was given by Cavendish in 1784. He performed numerous experiments on the explosion of mixtures of hydrogen and oxygen in varying proportions, and found that only when the proportions were two volumes of hydrogen to one volume of oxygen was the whole of the gas used up. If the gases were exploded in any other proportions, a residue of gas was left which was either hydrogen or oxygen.

Molecular Formulae of Hydrogen, Oxygen and Water. The explosion of hydrogen and oxygen in a eudiometer at the ordinary temperature does not give any information about the quantity of water produced from the volumes of the gases which combine. In order to obtain such information it is necessary to perform the experiment at a temperature well above 100° C. to maintain the water in the gaseous condition. A special form of eudiometer, provided with an outer vessel (Fig. 7) carrying the vapour of some liquid of higher boiling point than that of water, must be employed. Amyl alcohol, a liquid boiling at 132° C., is generally used for this purpose. The gases hydrogen, oxygen, and steam, must be measured under the same conditions as regards temperature and pressure, before and after explosion, since the volume of a gas depends upon physical conditions. For this purpose the gases are confined in the eudiometer over mercury.

In an actual experiment a dry mixture of two volumes of hydrogen and one of oxygen was introduced into the

closed limb of the apparatus shewn in the figure, and the
vapour of boiling amyl alcohol then passed through the
jacket. When the level of the mercury in the closed limb
no longer sank; that is, when the expansion of the gas due
to the heating ceased, the mercury was run off from the tap
until the two levels were the same, and the volume of gas

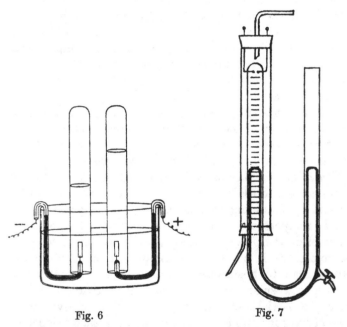

Fig. 6   Fig. 7

then read, and found to be 18 c.c. More mercury was then
run off to diminish the pressure, and therefore the violence
of the explosion—a very necessary precaution, otherwise
the apparatus would almost certainly fracture. The open
end of the apparatus was closed with the thumb and a
spark passed. On removing the thumb the mercury rose
in the closed limb. Mercury was poured into the open

limb until the two mercury levels were the same, when the volume was found to be 12 c.c. When the apparatus was allowed to cool, the mercury rose in the closed limb, almost filling it. Above the mercury a few drops of water and a small bubble of gas (due to error of experiment) were found.

Since the hydrogen and oxygen were previously mixed in the proportions of two volumes of the former to one volume of the latter: 12 c.c. of hydrogen combine with 6 c.c. of oxygen to form 12 c.c. of steam, all the volumes being measured under the same conditions of temperature, viz. 132° C. and barometric pressure. That is: *Two volumes of hydrogen combine with one volume of oxygen to form two volumes of steam.*

This result is simply a special case of Gay-Lussac's Law of Combination of Gases by Volume (1808). The law states that *gases combine in very simple proportions by volume, and the volume of the product, if gaseous, bears a simple relation to the volumes of the reacting gases.* Dalton (1808) made an unsuccessful attempt to explain this and similar results by assuming that equal volumes of gases at the same temperature and pressure contained the same number of *atoms*: an assumption which he found necessary to reject. It is not difficult to see that such reasoning would lead to absurdities, because if no distinction be drawn between atoms and molecules, it is obvious that two ultimate particles of water vapour must be derived from two of hydrogen and one of oxygen. This leads at once to the conclusion that one ultimate particle of steam is derived from one atom of hydrogen and half an atom of oxygen. As atoms are theoretically indivisible, Dalton's explanation was abandoned. A satisfactory hypothesis was however forthcoming when in 1811 Avogadro advanced the theorem that *equal volumes of all gases under the same physical conditions contain the same number of molecules.*

According to this reasoning, the formation of steam must take place by two molecules of hydrogen reacting with one molecule of oxygen yielding two molecules of water vapour. It follows that since one molecule of oxygen has yielded two molecules of steam, the molecule of oxygen has been divided into two parts, and therefore must consist of *at least* two atoms. It will be seen later that it is in the highest degree improbable that the oxygen molecule consists of *more* than two atoms. A study of the composition of hydrogen chloride (Chap. III) shews that the molecule of hydrogen is also diatomic. We may therefore express the synthesis of steam thus:

2 volumes of hydrogen + 1 volume of oxygen
= 2 volumes of steam.

Applying Avogadro's theorem:

2 molecules of hydrogen + 1 molecule of oxygen
= 2 molecules of steam.

Or, as an equation:

$$2H_2 + O_2 = 2H_2O.$$

An immediate deduction from Avogadro's hypothesis is that the molecular weights of gases are proportional to their relative densities. Taking the density of hydrogen as unity; since the molecule is diatomic, its molecular weight must of course be 2. As the density of oxygen is 16 the molecular weight of this gas is 32. The density of steam is found to be 9 relatively to hydrogen; the molecular weight is therefore 18. As there is at least one atom of oxygen in one molecule of steam, the value 18 for the molecular weight is obviously made up of 16 units due to the oxygen and 2 units due to the hydrogen. These figures lead directly to the representation of the molecules of hydrogen, oxygen, and steam by the formulae H_2, O_2, and H_2O respectively.

Empirical and Molecular Formulae. The molecular formulae of all gaseous substances rest upon Avogadro's hypothesis. The history of chemistry between 1811 and 1860 affords abundant proof of this. The hypothesis was published in 1811 but was at first largely neglected, with the result that there were no fewer than four 'systems' of chemistry between those years. The necessity for accepting Avogadro's reasoning was pointed out by Cannizzaro in 1860, and since then it has been received universally. Previous to that time there was endless confusion; several different formulae were in use for the simplest substances, and even when modern formulae were used it was due rather to chance than to reasoning.

The formula H_2O represents the molecule of steam. How then are we to represent the molecules of liquid water and of ice, which have obviously the same chemical composition as that of water vapour? Evidence can be adduced that the molecules of liquid and solid water are considerably more complex than those of steam; the degree of complexity is, however, uncertain. It would be correct, therefore, to express the molecular formulae of liquid water and of ice by $(H_2O)_n$ and $(H_2O)_{n'}$ in order to indicate that the molecular weights are unknown. Nevertheless the simple formula H_2O is frequently assigned to liquid and to solid water. In that sense it is an *empirical* formula; a formula, that is, which expresses the *relative* number of atoms in the molecule, but not necessarily the *actual* number.

Empirical formulae are constantly used to represent solid substances. For example, the formula $CaCO_3$ assigned to calcium carbonate expresses its chemical composition correctly. To represent the *molecule* of this substance the formula should strictly be written $(CaCO_3)_n$, where the value of n is still unknown.

In recent years much attention has been devoted to the

study of crystalline solids by the method of X-ray analysis devised by Sir W. H. Bragg and W. L. Bragg. Reliable information has been obtained regarding the arrangement of the atoms within the molecules of solids, and it appears that it will be possible to assign molecular formulae to substances the representation of which has hitherto been confined to empirical formulae.

Chemical Reactions of Water. As has already been mentioned, the solubility curves of some substances exhibit discontinuities due to the formation of hydrates; that is, substances which separate from aqueous solution with water of crystallization. These hydrates are definite compounds which are more or less readily dissociated by heat into the anhydrous substance and water vapour. Sometimes the colour of a hydrate is different from that of the anhydrous salt. Copper sulphate crystallizes from its saturated solution in water as a pentahydrate having the formula $CuSO_4.5H_2O$. If the finely powdered solid be heated for some time at a temperature of 100° C., four of the five molecules of water are lost, and the solid which remains has the formula $CuSO_4.H_2O$. This monohydrate is quite different in appearance from the pentahydrate; the former is of a pale bluish grey colour, the latter bright blue. If the monohydrate be heated to a temperature of about 250° C. it is converted into anhydrous copper sulphate which is colourless. Addition of water to the anhydrous salt or to the monohydrate results in the formation of the pentahydrate. The removal or addition of water is a reversible process.

Calcium sulphate is an interesting and important compound which forms hydrates; a dihydrate, $CaSO_4.2H_2O$, known as gypsum, and a hemihydrate, $2CaSO_4.H_2O$, which is produced by heating gypsum to a temperature of about 120° C. The hemihydrate, known as plaster of Paris, readily takes up water, forming the dihydrate, slightly expanding in the process and forming a hard mass due to the inter-locking of the crystals. If the dihydrate be heated much above 120° C., *all* the water of hydration is driven off and the resulting anhydrous calcium sulphate takes up water so slowly as to render it valueless either as a plaster or for taking casts. It is said to be *dead burnt*.

Deliquescence and Efflorescence. If ordinary washing soda (sodium carbonate decahydrate, $Na_2CO_3.10H_2O$) be freely exposed to the air the crystals lose their glassy appearance and gradually fall to an opaque powder, with a loss of more than half their weight. On the other hand, dry calcium chloride when exposed to the atmosphere rapidly absorbs moisture and becomes liquid. Salts which give up some or all of their water of hydration are said to be *efflorescent*, those which absorb atmospheric moisture, *hygroscopic*. If they actually liquefy they are termed *deliquescent*.

This is a case of equilibrium between the pressure of water vapour of the hydrate and that of the atmosphere. If the pressure of water vapour of a hydrated salt be greater than the pressure of water vapour present in the atmosphere, the salt will effloresce. On the other hand, if the salt has a lower pressure of water vapour, withdrawal of water from the atmosphere will take place. These processes continue until equilibrium is attained.

Many substances, particularly if in a fine state of division, readily 'adsorb' water. This is not in chemical combination and can be driven off by heat. In weighing fine powders it is necessary to remove this adsorbed moisture, either by the action of heat, or if the substance be of such a nature as to preclude heating, by exposure over a drying agent such as concentrated sulphuric acid, in a desiccator. The time required to dry a substance can be reduced considerably by using a *vacuum* desiccator; that is, one which can be exhausted of air by means of a pump.

Most substances are hygroscopic when finely divided, though by no means necessarily so when in fragments of larger size. Quartz, for instance, is sometimes used as a material for making exact weights because it does not tend to adsorb moisture to the same extent as glass, though in a fine state of division it is hygroscopic.

Hydrolysis. When some salts are dissolved in water they may be recovered by the simple process of evaporating to dryness. Sodium chloride is a case in point, but many salts when once dissolved in water cannot be again recovered unchanged by evaporation. For instance, when a solution of ferric chloride is evaporated to dryness a residue remains which will not again dissolve in water completely. The ferric chloride and the water interact, not by simple addition, but in a manner which may be represented by the equation:

$$FeCl_3 + 3H_2O = Fe(OH)_3 + 3HCl.$$

During the evaporation both water and hydrochloric acid escape, and the residue consists of both ferric chloride and ferric hydroxide, the latter being insoluble in water. Such a process is termed *hydrolysis*, which may be defined as any reaction in which water takes an essential part, indicated by a chemical equation, other than by simple addition.

Action of Metals on Water. The Preparation and Properties of Hydrogen. Certain metals react with water at ordinary temperatures with the formation of soluble hydroxides and the evolution of hydrogen. If, for example, a small piece of potassium, or sodium, or calcium be placed in water a reaction takes place, very energetically with potassium, less so with sodium, and gently with calcium. In the latter two cases the evolved gas may be collected without difficulty,

but with potassium the reaction is too violent for convenience. In all three cases the remaining liquid is strongly
alkaline, and on evaporation (preferably in a silver or
nickel dish) leaves a colourless solid behind—the metallic
hydroxide. The reactions are represented thus:

$$2K + 2H_2O = 2KOH + H_2,$$
$$2Na + 2H_2O = 2NaOH + H_2,$$
$$Ca + 2H_2O = Ca(OH)_2 + H_2,$$

only one-half of the hydrogen being displaced from the
water molecule in each instance.

When it is desired to prepare hydrogen by this reaction,
sodium is used in the form of an alloy with mercury—
sodium amalgam. The reaction is then quite gentle, the
mercury acting as a diluent but taking no other part in
the change. As the gas is insoluble in water it may be
collected without loss.

Some metals, such as iron and magnesium, have little or
no action upon water at ordinary temperatures, but when
heated in a current of steam displace the hydrogen from
the water completely, forming the metallic oxide. Thus if
a current of steam be passed over iron heated to bright
redness, hydrogen is liberated and an oxide approximating
in composition to magnetite, Fe_3O_4, is formed in accordance with the equation:

$$3Fe + 4H_2O = Fe_3O_4 + 4H_2.$$

This reaction has been used for the manufacture of
hydrogen on a large scale. It has also been employed for
the preparation of a non-rusting iron or steel. The process
consists in heating bright steel articles in a gentle current
of steam free from air, when the steel becomes covered with
a thin coherent film (or patina) of the bluish coloured oxide
which protects it from atmospheric corrosion. This reaction

is reversible; if hydrogen be passed over heated iron oxide, water and metallic iron are obtained.

If a strip of burning magnesium ribbon be plunged into the steam issuing from water boiling in a flask, it will continue to burn, forming magnesium oxide and liberating hydrogen, which of course also burns at the mouth of the flask.

$$Mg + H_2O = MgO + H_2.$$

It must not be supposed that *all* metals which can be oxidized at high temperatures are necessarily capable of decomposing steam like iron and magnesium. Metallic copper, for instance, is not oxidized by steam even at a red heat.

If steam be heated to a sufficiently high temperature it is in part resolved into its elements, or 'dissociated.' The term *dissociation* was originally applied to reversible decompositions in which one at least of the products of decomposition was gaseous, but now has a wider meaning and includes all reversible reactions in which complex molecules are resolved into simpler ones. The dissociation of steam may be shewn by passing it through a white-hot platinum or porcelain tube and collecting the gases produced over water, the air having been previously removed from the apparatus by allowing steam to pass through the tube for some time before heating it. The same result may be obtained by fixing a coil of platinum wire in the neck of the flask of a simple distilling apparatus by means of copper rods passing through the stopper (Fig. 8), attaching the rods to a suitable source of electricity by which the platinum wire can be heated to bright redness, steaming out all air from the apparatus, and then switching on the current. When a small quantity of gas has been driven over by the steam into the collecting tube its explosive nature can be demonstrated.

The platinum (or porcelain) used in these experiments is not oxidized. The effect is a thermal dissociation simply, and under the conditions described only a very small proportion of the steam undergoes this change. Langmuir estimates the amount of dissociation at less than one-half per cent. at the temperature at which platinum begins to melt, and not more than 12 per cent. at 3000° C., the temperature of the electric arc.

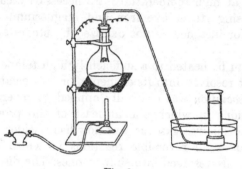

Fig. 8

Small quantities of hydrogen for experimental purposes may be conveniently prepared by the action of sulphuric or hydrochloric acid on certain metals. Thus, if granulated zinc be placed in dilute sulphuric or hydrochloric acid a copious evolution of the gas will follow, under certain conditions, without heating. Pure dilute sulphuric acid has little or no action on pure zinc. A voltaic circuit is necessary, and this is usually established by impurities in the zinc itself. If such impurities are not present, energetic action can be induced by various means, such as (a) placing scraps of metallic platinum in contact with the zinc, (b) adding a few drops of a solution of chloroplatinic acid or of copper sulphate, when the zinc becomes covered with a thin film of metallic platinum or copper, and the voltaic circuit is established. A solution of zinc sulphate is produced with evolution of gas according to the equation:

$$Zn + H_2SO_4 = ZnSO_4 + H_2.$$

Iron or magnesium may be substituted for the zinc. In the case of iron the action is less energetic; with magnesium it is quite violent, unless the acid is very dilute. With hydrochloric acid either magnesium, zinc, iron, or tin may be used. The metals are written in descending order of the energy of the reaction, and in the case of tin fairly concentrated acid will be required, and also the application of external heat to start the action.

It should be noted that concentrated sulphuric acid rarely attacks metals in the cold, but it frequently does so on heating. Hydrogen is never evolved in such cases; only *dilute* sulphuric acid can be used for the preparation of this gas. Hydrochloric acid, on the other hand, invariably evolves hydrogen when it reacts with a metal, whatever its concentration.

The cheap preparation of hydrogen on a large scale has in recent years become important for various technical applications. The filling of military airships in the field demands a rapid preparation from materials which are neither heavy nor dangerous. For such purposes 'hydrolith' and 'ferrosilicon' are materials which have been employed. Hydrolith contains about 90 per cent. of calcium hydride, CaH_2, and reacts with water as follows:

$$CaH_2 + 2H_2O = Ca(OH)_2 + 2H_2.$$

According to this equation, one gramme of hydrolith yields approximately one litre of hydrogen.

Ferrosilicon is largely uncombined silicon, and reacts

with dilute sodium hydroxide (better in the presence of calcium hydroxide) thus:

$$Si + 2NaOH + H_2O = Na_2SiO_3 + 2H_2,$$

the commercial material giving a yield of hydrogen similar to that from hydrolith.

Hydrogen for military balloons has also been prepared by the action of caustic soda on aluminium as described below. The only materials requiring transport for this method being the light metal aluminium, and solid sodium hydroxide, the water of course being obtained locally.

It is sometimes necessary to prepare hydrogen in alkaline solution, i.e. where the presence of acids is inadmissible, as in Fleitmann's test for arsenic (Chap. VIII), although it is rarely needful to collect the gas itself under these conditions. A convenient method for doing this is found in the reaction between sodium hydroxide solution and metallic aluminium or zinc. The former metal reacts very energetically with caustic soda solution, liberating hydrogen:

$$2Al + 2NaOH + 2H_2O = 2NaAlO_2 + 3H_2.$$

The reaction between soda and zinc is so slow that in most cases it is necessary to accelerate it by producing a voltaic circuit. The so-called zinc-copper couple generally used is made by covering granulated zinc with a dilute (3 per cent.) solution of copper sulphate, allowing to stand for a short time, pouring off the liquid and washing the black copper-coated zinc several times with water.

$$Zn + 2NaOH = Na_2ZnO_2 + H_2.$$

The manufacture of margarine from certain vegetable oils entails a process known as 'fat hardening,' i.e. combining the oils with hydrogen. For this purpose the 'hydrogenizing' gas is produced electrolytically, finely divided nickel being used as a 'catalyst' (p. 70) to accelerate the union of the hydrogen and the oil.

Hydrogen is now manufactured on a very large scale for the synthetic production of ammonia (Haber's process, p. 144). Formerly it was used with oxygen in the oxy-hydrogen flame, which has a temperature of over 2000° C. This flame, impinging on a piece of quicklime, renders it incandescent—the so-called limelight. For the production of high temperatures, however, hydrogen is now largely superseded by acetylene, as the latter gives a still hotter flame.

Over a wide range of temperature and pressure hydrogen is a gas; odourless, colourless and tasteless. At very low temperatures it may be condensed to a colourless liquid, and finally to a solid. The critical temperature is − 235° C. (Olszewski, 1895). The lowness of this was not realized by the early experimenters on the liquefaction of gases, hence the reason why hydrogen resisted liquefaction for so many years. It is the lightest known substance, having only 1/14·4 of the density of air, and this renders it valuable for aeronautical purposes. In this connection the inflammability of the gas is a serious disadvantage however, and wherever possible it is being replaced by the much more costly gas helium. The latter, although twice as heavy as hydrogen, has practically the same lifting power, and is not inflammable.

As has been shewn, hydrogen and oxygen combine with violence under the influence of the electric spark. The same result follows if finely divided platinum be introduced into the mixture. This can be shewn experimentally by putting a few cubic centimetres of dilute chloroplatinic acid solution into a stout-walled soda water bottle and passing hydrogen through the solution. Metallic platinum is first produced, and when sufficient hydrogen has collected in the bottle to form an explosive mixture with the air present, detonation takes place.

If a mixture of oxygen and hydrogen be passed slowly over various solids, such as porous porcelain, metallic gauzes, etc., heated to a moderate temperature (below a red heat) combination takes place gently and without anything in the nature of an explosion.

Hydrogen combines directly with most non-metallic elements and also with some metals. The metallic hydrides as a class are not remarkable for their stability. Most of them react readily with water, with evolution of hydrogen, while non-metallic hydrides are much more stable. Some importance was formerly attached to this as a means of distinguishing between metals and non-metals, but it is now recognized that no conclusions of value can be deduced from this difference in behaviour.

Hydrogen is the essential constituent of all acids. An acid is sometimes defined as a compound which contains hydrogen replaceable by the metal of a soluble metallic hydroxide under all conditions of dilution. No definition of an acid which is altogether satisfactory can be given. The subject is again referred to in Chap. x.

The Gravimetric Composition of Water. The composition of water by weight can be deduced from the known densities of its constituent gases. The gravimetric ratio was determined directly by Dumas in 1842. He prepared hydrogen by the action of zinc on dilute sulphuric acid, and purified it from the hydrides of sulphur and arsenic by passing the gas through solutions of lead acetate and silver sulphate. The gas was then passed over caustic potash to remove acid spray, and finally over phosphorus pentoxide to dry it thoroughly. The hydrogen thus purified and dried was passed through a tube containing copper oxide heated to redness. The tube and its contents were carefully weighed before and after the experiment: the loss in weight representing the weight of oxygen supplied by the copper oxide. The bulk of the water produced was collected in a bulb, and the remainder removed from the excess of hydrogen used by passing the issuing gas through weighed tubes containing phosphorus pentoxide. Thus the weights of oxygen used and of the water formed being known, the weight of hydrogen was also known. A simplified apparatus is represented in Fig. 9. One of Dumas' experiments gave the following numbers: 67·282 grm. of water were produced from 59·788 grm. of oxygen, and therefore 59·788 grm. of oxygen combine with 7·494 grm. of

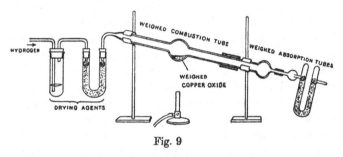

Fig. 9

hydrogen to form 67·282 grm. of water. The mean result of a large number of experiments shewed that 1 grm. of hydrogen combines with 7·98 grm. of oxygen.

The combining ratios of hydrogen and oxygen are therefore very nearly 1 to 8. This is sometimes expressed by saying that the equivalent weight of oxygen is 8. The equivalent weight of an element is usually defined as that weight of it which will combine with or take the place of unit weight of hydrogen.

The deviation from the ratio 1 : 8 shewn by Dumas' figures is real, and cannot be attributed to experimental error. Later investigators are all agreed that the ratio is greater than 1 : 8. Scott (1893), who prepared his hydrogen by the action of steam upon sodium and determined the combining volumes, arrived at the ratio 1 : 7·931. Morley (1895) found the ratio to be 1 : 7·9395, using hydrogen prepared electrolytically from alkaline solution. The latter weighed his hydrogen directly, by absorbing the purified gas in palladium. This metal when cold absorbs over 900 times its own volume of hydrogen, yielding it again on heating.

The discrepancies between these numbers for the combining ratio of hydrogen and oxygen though small are important in equivalent weight determinations; so important indeed that the theoretical unit, hydrogen, has been abandoned in all modern work in favour of a new unit, obtained by arbitrarily fixing the equivalent weight of oxygen at 8. Other equivalent weights are determined with reference to this oxygen standard, and the definition becomes: *The equivalent of an element is that weight of it which combines with or replaces eight parts by weight of oxygen*, and to this is usually added, *or that weight of any other element which in turn combines with or replaces eight parts by weight of oxygen*, since the determination is sometimes made indirectly through another element, for experimental reasons. For instance, although sodium combines with oxygen to form an oxide, experiments on the oxide do not give results which are accurate or concordant, as the product is not a single chemical individual but a mixture of oxides. Much more reliable numbers are obtained by determining the weight of sodium which combines with one equivalent weight of chlorine, and since the latter equivalent is very well known, the oxygen equivalent can be calculated.

The atomic weights of all elements, with the exception of the gases of the helium group, are derived from equivalent weight determinations, and practical considerations have led to the choice of oxygen as the standard. Most elements form well-defined compounds with oxygen which are suitable for accurate analysis, but this is rarely the case with hydrogen compounds. It was formerly the custom to publish *two* series of atomic weights, having hydrogen and oxygen as the respective standards. Inspection of the two sets of figures shews differences which are of importance only where great accuracy is required: for example, the atomic weight of calcium is 40·1 on the oxygen standard, and 39·8 when referred to hydrogen. For ordinary analytical work it is immaterial which value be accepted, and actually the value 40 is used. At the present time atomic weights are always expressed in terms of the oxygen standard.

The Hydrogen Equivalent of Metals. The equivalent weights of those metals which evolve hydrogen by their action upon hydrochloric or dilute sulphuric acid may be determined by collecting the gas obtained from the action of excess of the acid upon a known weight of the metal. The most convenient form of apparatus is that due to Ostwald (Fig. 10). A weighed piece of zinc, not exceeding 0·5 grm., is placed in the flask and covered with about 25 c.c. of water. The small test tube contains about 5 c.c. of concentrated sulphuric or hydrochloric acid. The joints are then made air-tight and the level of the water in the limbs A and B adjusted. The level in A is read (no attempt being made to bring the reading to zero) when the whole apparatus has assumed the temperature of the room, which is shewn by the water levels remaining constant. The

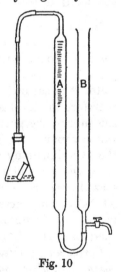

Fig. 10

temperature is carefully noted. The concentrated acid is then mixed with the water by tilting the flask; the action being accelerated if necessary by the application of heat. When all the metal has disappeared the water levels are again adjusted and the apparatus allowed to regain the temperature of the room, this being indicated as before by the constancy of the water levels. It should be noted that accuracy depends upon the room temperature remaining unchanged throughout the duration of the experiment, since there is no mechanism for correcting any change of temperature at this stage, the volume of the flask being unknown. In an actual experiment the following results were obtained:

0·336 grm. of zinc displaced 126 c.c. of hydrogen measured at 14° C. and 741 mm. pressure.

In order to calculate the weight of hydrogen obtained in this experiment the volume of the gas collected must be corrected to standard conditions of temperature and pressure (0° C., and 760 mm.). As the gas was collected over water it is saturated with water vapour (see p. 19), and therefore the total pressure (= barometric height) is really the sum of the partial pressures of the hydrogen and the water vapour. The latter is found for the temperature of the experiment by reference to the table of vapour pressures (p. 20) and this must be subtracted from the observed pressure in order to obtain that of hydrogen alone. At 14° C. the pressure of water vapour is 12 mm. (the nearest whole number). The pressure of the hydrogen is therefore 741 − 12 mm. = 729 mm.

Correction for Pressure. Boyle's Law (1662) states that *the volume of a given mass of gas is inversely proportional to the pressure upon it when the temperature is constant,* or, if *P* and *V* are the pressure and volume of a given mass of gas as observed in the experiment, and P' and V' the corresponding values for standard pressure,

$$PV = P'V',$$

$$760V' = 729 \times 126 \text{ c.c.},$$

or $$V' = 121 \text{ c.c.}$$

Correction for Temperature. Charles's Law (1787) states that *Gases expand or contract by 1/273 of their volume at 0° C. for every degree centigrade through which they are heated or cooled, the pressure remaining constant.* An obvious deduction from this law, assuming that it remains valid at all temperatures, is the conception of an absolute zero, since, if a gas could be cooled to $-273°$ C., its volume would be *nil*; that is, a destruction of matter would have been accomplished. This temperature has been closely approached but never actually reached, and represents an absolute cessation of the vibratory motion which constitutes heat: a greater degree of cold cannot even be imagined. The absolute zero of temperature, or zero on the absolute scale, is therefore $-273°$ C. There are other lines of investigation which indicate this same zero of temperature. For instance, the electrical resistance of pure metals diminishes on cooling in a manner which indicates perfect conductivity, or, in other words, the absence of all electrical resistance, at $-273°$ C.

The law of Charles may therefore be alternatively stated: *The volume of a given mass of gas at constant pressure is directly proportional to the temperature on the absolute scale.* Temperatures on the absolute scale are obtained by adding 273 to the reading on the centigrade scale.

The correction for temperature may be expressed by the equation:

$$\frac{v}{T} = \frac{v'}{T'},$$

where v and T denote the observed volume and absolute temperature and v' and T' the corresponding values for standard temperature, $0°$ C.

$$\frac{v'}{273} = \frac{121}{273 + 14}; \text{ and } v' = 115 \text{ c.c.}$$

It is more convenient to apply the corrections simultaneously, thus:

$$\frac{PV}{T} = \frac{P'V'}{T'},$$

$$\frac{760V'}{273} = \frac{(741 - 12) \times 126}{273 + 14},$$

from which $V' = 115 \text{ c.c.}$

It has been found that 1 c.c. of hydrogen at $0°$ C. and 760 mm. pressure weighs very accurately 0·00009 grm. Another way of expressing this is that the volume of 1 grm. of hydrogen at normal temperature and pressure (N.T.P.) is 11·11 litres. The double value, 22·22 litres, is called the gramme molecular volume, since one molecular weight of any gas, expressed in grammes, occupies 22·22 litres at N.T.P.

The weight of 115 c.c. of hydrogen at N.T.P. is equal to 0·00009 × 115 grm. Therefore the equivalent weight of zinc becomes:

$$\text{Eq. Wt.} = \frac{0·336}{0·00009 \times 115} = 32·5.$$

The relation between equivalent weight and atomic weight may be expressed by the equation:

Atomic weight = Equivalent weight × valency,

the valency being a small integral number. Valency is not determined directly, but the discussion of this subject is deferred.

Oxygen, Preparation and Properties. The evolution of oxygen at the anode in the electrolysis of dilute sulphuric acid has already been mentioned. The gas may be obtained by heating many of its compounds, and since oxygen is very sparingly soluble in water, it can be collected by displacement of water.

Priestley (1774) obtained oxygen by heating mercuric oxide (HgO) which dissociates into its elements mercury and oxygen at a temperature a little above the boiling point of mercury. This oxide had previously been obtained by heating the metal in air at a temperature somewhat below its boiling point. The ultimate source of the oxygen was therefore the atmosphere.

$$2HgO \rightleftharpoons 2Hg + O_2.$$

Lead oxide, PbO, on being heated in contact with air also takes up oxygen, forming red lead, Pb_3O_4, which in turn, on being heated more strongly, yields up part of its oxygen, becoming once more lead monoxide,

$$6PbO + O_2 \rightleftharpoons 2Pb_3O_4.$$

A more convenient method of obtaining oxygen from the atmosphere, worked for many years but now obsolete, was known as Brin's process. When barium oxide, BaO, is maintained at a temperature of about 500° C. in contact with dry air freed from carbon dioxide, it unites with oxygen to form the dioxide or peroxide BaO_2. This reaction reverses at a temperature of 800° C.

$$2BaO + O_2 \underset{\text{at } 800° \text{ C.}}{\overset{\text{at } 500° \text{ C.}}{\rightleftharpoons}} 2BaO_2.$$

In the operation it was soon found unnecessary to continue the wasteful and troublesome process of changing the temperature of the large containers, since the reaction could be far more easily and economically controlled and reversed by changes of pressure. This is an application of Le Chatelier's principle; increase of pressure favours the formation of the peroxide, while lowering the pressure assists the dissociation of the peroxide into oxygen and barium monoxide.

The large vessels filled with barium oxide were maintained at a constant temperature of about 500° C., and air was forced in under pressure, there being, of course, a mechanism by which the spent air, robbed of the greater part of its oxygen, could escape. When the percentage of barium peroxide had reached its most advantageous proportion the operation was reversed. Oxygen was removed by means of a pump, and compressed into cylinders. The purity of the product varied between 93 and 96 per cent. of oxygen, the cost being far below that of the gas made by any laboratory method.

By far the most economical method of obtaining oxygen on a large scale is the distillation of liquid air, and this has quite superseded Brin's process. The principle involved in the liquefaction of air is simply that when gases are compressed heat is evolved; when they are allowed to expand heat is absorbed.

Air, freed as far as possible from water vapour and carbon dioxide by being made to pass through towers filled with potash and quicklime, is pumped into a vertical steel cylinder at a pressure of 200 atmospheres. Considerable heat is evolved in the compression, and therefore the gas is made to pass through a spiral tube immersed in cold water before it enters the cylinder. From the cylinder the compressed air passes into a very long copper tube of small bore (about $\frac{1}{8}$ inch) wound spirally in the liquefying chamber, and escapes through a minute orifice at the end of the tube. On escaping it expands and its temperature

falls. The cooled gas passes over the outside of the copper
spiral and still further reduces the temperature of the compressed gas within, the effect being cumulative. When this
process has continued for some time, the temperature of
the air at the moment of its escape and expansion from the
small orifice is so low that some of it is liquefied.

When liquid air evaporates, the nitrogen, having the
lower boiling point, comes off more freely than the oxygen,
and the last portions to evaporate are nearly pure oxygen.
The gas which is prepared in this way for the market contains but a small percentage of impurity, chiefly nitrogen.
A ready method of preparing small quantities of oxygen
for laboratory experiments is found in the heating of
potassium chlorate, $KClO_3$. This compound, which does
not contain crystal water, on heating below a red heat
melts without evolution of gas. At a higher temperature
it decomposes, giving up oxygen and leaving potassium
chloride, KCl, thus:

$$2KClO_3 = 2KCl + 3O_2,$$

but this only represents the final stage. An intermediate
product, potassium perchlorate, $KClO_4$, is formed, though
this is finally resolved into potassium chloride and oxygen
if the heating be prolonged and the temperature high
enough. The preparation of potassium perchlorate from
potassium chlorate can easily be made by melting the
latter, and continuing the heating until the molten mass
becomes somewhat viscid. On cooling and crystallizing
the residue, first from hydrochloric acid which decomposes
the remaining chlorate, and finally from hot water, crystals
of the perchlorate in a high state of purity may be obtained.

If a few grammes of potassium chlorate be melted in a
test tube, and when the mass is liquid and tranquil a little
manganese dioxide, MnO_2, be added, a violent evolution of

oxygen is observed. The action of the manganese dioxide is peculiar. Apparently no chemical change has taken place between the chlorate and dioxide, because if the cooled contents of the tube be extracted with water the manganese dioxide, being insoluble, can be recovered unchanged both in properties and mass. Such an action is called *catalysis*, the dioxide in this instance being the *catalyst*. The latter is defined as a substance which assists a chemical change without itself undergoing any chemical change. If the dioxide has undergone any such change (and on this point the evidence is not altogether conclusive) it must have been a *cyclic* one, i.e. the compound ends up precisely as it began, and therefore can neither have added nor subtracted energy in the reaction. Catalysis has in recent years developed an enormous importance, especially in chemical manufactures. The subject is again referred to later.

The laboratory preparation of oxygen from potassium chlorate is best carried out by grinding up the chlorate with about one-fifth of its weight of manganese dioxide and heating this mixture, preferably in a round bottomed flask of hard glass. Only a very moderate temperature is required. The gas comes off very freely and may be collected over water in the usual way.

The most striking property of oxygen is its capacity for combining with other elements, and indeed the energy of such reactions is one of its distinguishing features. Substances which burn in air do so more vigorously in pure oxygen, since the diluting action of the nitrogen, which constitutes four-fifths of the atmosphere, is absent. The burning of hydrogen in oxygen or air is well known, and can be shewn to be a reciprocal process. In general terms, if a gas A burns in a gas B, B can be made to burn in A. The apparatus for shewing this is illustrated in Fig. 11.

The artificial production of heat is nearly always oxidation, the oxidized substances being usually coal, coke, wood, oil, coal gas or other gaseous fuel. In fact, other sources of heat, such as the radiant heat of the sun and electricity derived from water power, receive little application for technical purposes as compared with the burning

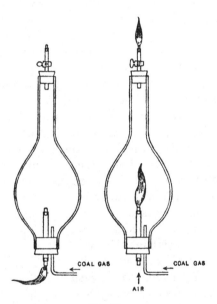

Fig. 11

of various forms of fuel. *Combustion* is a vague term which usually implies any chemical change accompanied by a rise in temperature sufficient to cause visible heat, that is, light: no precise definition can be given, but usually union with oxygen is understood. The absorption of oxygen by the blood and its subsequent use in oxidizing the tissues of the body is sometimes quoted as an instance of 'slow combustion.' It may be mentioned here that the quantity of oxygen absorbed by blood is not proportional to the pressure of the oxygen, and therefore this is not a case of simple solution.

Combustion in oxygen would seem to be a more complex phenomenon than the direct union of a substance with the gas. Dixon has shewn that carbon monoxide, CO, which ordinarily burns readily with oxygen forming carbon dioxide, will not do so if the gases are exhaustively dried. Baker has shewn many other instances of the same phenomenon. Phosphorus was distilled in oxygen without any chemical change. A mixture of hydrogen and oxygen was heated to the melting-point of silver without explosion. In all such cases the gases used were previously dried most carefully with phosphorus pentoxide. Water vapour would seem to be a catalyst which is necessary if combustion is to take place. This is certainly true in some cases.

Ozone. When a slow stream of oxygen is passed through an annular space between two tubes (Fig. 12) which are insulated from each other and maintained at a high difference of potential, in such a way that there is a leakage of electricity without actual sparking, the oxygen undergoes a remarkable change and a characteristic odour is noticed. The issuing gas will attack many substances towards which oxygen is inert at ordinary temperatures. For example, mercury quickly begins to 'tail,' that is, to adhere to the sides of the bottle, when exposed to the gas; indigo is bleached,

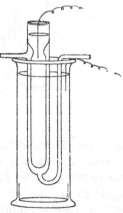

Fig. 12

and a solution of potassium iodide becomes brown from the liberation of iodine. Clearly, then, the so-called 'silent' electric discharge has increased the chemical energy of the oxygen. If the gas so altered in properties be passed through a hot tube of some inert material such as porcelain, a reverse change takes place. The new properties are lost, and the issuing gas is once more ordinary oxygen, equal in volume and mass to the original oxygen. It is plain, therefore, that the new substance produced by the electric stress and known as ozone consists entirely of oxygen atoms. Since ozone and oxygen differ in properties, their molecular structures must differ. The only way in which this is possible, since but one kind of atom is present, is *either* in the number of atoms in their respective molecules, that is, their *atomicity*, or in their intra-molecular arrangement, which is the relationship of the atoms to each other *within* the molecule.

This phenomenon of the existence of an element in more than one form (the various forms, of course, differing in specific properties) is known as *allotropy*. The phenomenon is of fairly common occurrence, particularly in the case of solid elements. Among gases it is rather exceptional, the only well-known example, in addition to oxygen, being the element nitrogen.

An essential difference between the various allotropes of an element lies in the energy content. There is a difference between the quantities of heat evolved when the same mass of each of the allotropes of carbon is burnt completely in oxygen, although the product is the same, both in properties and quantity. Ozone is more energetic than oxygen.

Small quantities of ozone are produced when phosphorus slowly oxidizes in damp air. Traces of the gas may be detected in the oxygen evolved by the electrolysis of dilute sulphuric acid. The smell of ozone is noticeable during the working of electric machines, especially when sparks are produced: lightning flashes are sometimes held to account for the presence of traces of ozone in the atmosphere.

Many of the tests for ozone are unsatisfactory. Ozone paper, which is filter paper soaked in a solution of potassium iodide and starch and then dried, turns blue in presence of traces of ozone, but this colour is also given by other substances often present in the atmosphere, such as the higher oxides of nitrogen. The test therefore is useless unless the absence of such substances is assured. 'Tetramethyl base' is much more reliable, since it produces different colours with the substances which turn potassium iodide-starch papers blue. It gives a violet colour with ozone, blue with chlorine or bromine, straw yellow with nitrous fumes, and no colour with hydrogen peroxide.

The assertion that ozone is present in the air, especially near the sea, is based almost entirely upon the behaviour of starch-potassium iodide paper, and therefore is untrustworthy. It now seems probable that ozone, if present in the atmosphere at all, cannot exceed one part in half a million parts of air.

The proportion of oxygen converted into ozone by any form of ozonizing apparatus is never very high, whether pure oxygen or air be used. The most careful ozonizing of oxygen probably never exceeds 25 per cent. of conversion, and under ordinary conditions the proportion is very much smaller.

Ozone has some technical applications due to its powerful oxidizing properties. It has been used for sterilizing water, and also for purifying air in enclosed spaces such as underground railways, cold storage rooms, etc.

The Molecular Formula of Ozone. Since ozone cannot be obtained in pure condition or collected over mercury, measurements of its volume or density must be made indirectly. When oxygen is ozonized there is a contraction in volume, and hence, in accordance with Avogadro's theorem, a diminution in the number of molecules. It follows that the ozone molecule contains a larger number of atoms than the oxygen molecule. Schönbein observed that ozone is absorbed by certain essential oils, such as oil of turpentine, and he made use of this property of oil of turpentine to determine the percentage of ozone present in ozonized air or oxygen. When ozone is heated it reverts to ordinary oxygen with an increase in volume. Soret found that for a given specimen of ozonized oxygen, the volume absorbed by oil of turpentine was double the increase in volume obtained by heating the gas. Thus if 100 c.c. of ozonized oxygen diminish by absorption to $100 - 2x$ c.c., and in another experiment on the same sample, 100 c.c. on heating become $100 + x$ c.c., it is clear that $2x$ c.c. of ozone yield $3x$ c.c. of oxygen. By the application of Avogadro's theorem the relationship between oxygen and ozone must be represented by the equation:

$$2O_x \rightleftarrows 3O_2.$$

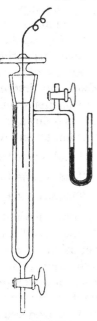

Fig. 13

Thus, if the diatomicity of the oxygen molecule be accepted (and reasons for this have already been given) the ozone molecule must be regarded as triatomic, and represented by the formula O_3.

The atomicity of the ozone molecule may be demonstrated by means of the apparatus in Fig. 13, which obviates the necessity of using pure ozone or of collecting ozonized oxygen over a liquid. The apparatus is essentially a Siemens ozonizer in which the inner tube is movable by means of a handle, while between the two tubes a thin glass capillary is gently held between glass projections on the tubes themselves. The sealed capillary holds a small quantity of oil of turpentine. Oxygen is passed through the apparatus and the taps closed. The manometer, containing a little coloured water, is adjusted by momentarily opening one of the taps. The electric stress is applied for a few minutes and the apparatus afterwards allowed to regain the temperature of the room. The contraction shewn by the manometer is noted. A sharp turn of the handle breaks the capillary, when the ozone is dissolved by the liberated turpentine and a further contraction takes place. The essence of this experiment is that *the second contraction is double the first one.*

If the first contraction be n units and the second $2n$ units, it follows, since the contractions are proportional to the volumes removed:

Total oxygen removed $= 2n + n = 3n$ c.c.

Total ozone formed $= 2n$ c.c.

Therefore $3O_2 = 2O_x$, and $x = 3$.

Another method for determining the molecular formula of ozone, also due to Soret, depends on the rate of diffusion (or effusion) of gases. It is well known that if two gases be mixed there is no tendency towards separation or segregation, however greatly they may differ in density. Such a mixture remains homogeneous indefinitely, and indeed may be considered as a solution of one gas in another. All gases are infinitely soluble in each other, or, in other words, form homogeneous mixtures in all proportions. The atmosphere consists mainly of oxygen and nitrogen in the proportions of one-fifth of the former to four-fifths of the latter, this proportion being practically constant throughout the world. Careful analyses by Frankland of air taken at Chamonix and at the summit of Mont Blanc shewed no variation in the proportion of the two gases. If a mixture of two gases which differ in density be enclosed in a vessel of porous material, such as unglazed porcelain, the lighter gas will diffuse through the walls of the vessel more rapidly than the heavier one. Many interesting experiments can be devised to illustrate this phenomenon. For example, if a mixture of two volumes of hydrogen and one volume of oxygen be passed very slowly through a long churchwarden tobacco pipe attached to an apparatus for collecting the gas, it will be found that the latter will not explode on the application of a light; on the other hand, it may be sufficiently rich in oxygen to rekindle a glowing splinter of wood. The light gas hydrogen has diffused through the walls of the pipe much more rapidly than the heavier oxygen. Graham (1831) investigated this phenomenon quantitatively, and shewed that *The relative rates of diffusion of gases when under the same conditions are inversely proportional to the square roots of their densities.* This is known as Graham's Diffusion Law. For example:

Density of hydrogen 1, density of oxygen 16. The rates of diffusion of the two gases are therefore

$$\frac{1}{\sqrt{1}} : \frac{1}{\sqrt{16}}, \; = 1 : \tfrac{1}{4}; \; = 4 : 1.$$

That is, hydrogen diffuses four times as rapidly as oxygen, under the same conditions.

It should be noted that the porous material forms no part of diffusion phenomena. While allowing diffusion to take place through it (more slowly perhaps) it hinders mechanical mixing by convection currents. If two gas jars be placed mouth to mouth, the lower containing the heavy gas carbon dioxide and the upper containing hydrogen, the two gases diffuse into each other in spite of their great difference in density, ultimately forming a homogeneous mixture.

The accepted explanation of diffusion and allied phenomena is to be found in the kinetic theory of gases. According to this, the molecules of a gas are in constant and rapid motion. The free path of a gaseous molecule is a straight line between two collisions, and is large in comparison with the diameter of the molecule. The pressure of a gas is considered to be due to the bombardment of the walls of the containing vessel by the molecules.

Let us consider two similar vessels containing hydrogen and oxygen respectively at the same temperature and pressure. The total pressure on the walls of each of these vessels must be the same, and therefore the kinetic energy of each of the two gases must also be equal. Since the mass of the oxygen molecule is 16 times that of the hydrogen molecule, and since the kinetic energy of a mass M moving with a velocity V is represented by $\frac{1}{2}MV^2$, it follows that the mean velocities of the two gases are inversely proportional to the square roots of their relative masses (densities) under the same conditions of temperature and pressure. In other words, the mean velocity of the hydrogen molecule is four times that of the oxygen molecule. This reasoning applies to all gases. The actual velocities of gaseous molecules are very considerable. The mean velocity of the hydrogen molecule under standard conditions of temperature and pressure has been calculated to be approximately 1850 metres per second. It is not surprising that molecules travelling with such velocity should be able to traverse the constricted channels of a porous diaphragm, and obviously the greater the velocity the greater will be the number of molecules which will pass through in a given time. The phenomena of diffusion may be easily visualized from such considerations.

Soret's method of procedure was to compare the rate of diffusion of ozone with that of chlorine under identical conditions. The two gases, both diluted with oxygen to the same degree, were made to diffuse separately through the same apparatus into solutions of potassium iodide. From the weights of iodine set free, the relative masses of chlorine and ozone which had diffused through the diaphragm could be calculated, since the weight of iodine which is liberated by a given mass of each of these gases is well known. It was found that 5·95 grm. of ozone diffused through the apparatus in the same time as 4·90 grm. of chlorine. Since the density of chlorine is 35·5, it follows that the density of ozone is very nearly 24, the molecular weight being therefore 48, which corresponds with a tri-atomic molecule, O_3.

The transformation of oxygen into ozone is an endo-thermic process. According to Le Chatelier's theorem, it would be expected that the yield of ozone should be in-creased by subjecting oxygen to the silent discharge at high temperatures. This is not what is observed, however. Indeed, it would seem to be inconsistent with the familiar experiment of decomposing ozone by passing the gas through a heated tube. It must be noted that the effect of heat upon chemical change is twofold; firstly upon chemical equilibrium, and secondly upon reaction velocity. The former follows Le Chatelier's theorem; the latter is always accelerated by raising the temperature. As a general rule, a rise of 10° C. doubles the velocity of a chemical reaction. If the ozonization is effected at temperatures above the ordinary, the increased yield is more than lost by the increased rate at which the ozone is decomposed.

Hydrogen Peroxide. If a hydrogen flame be rapidly

cooled, the water produced possesses oxidizing properties. This may be shewn by allowing a jet of burning hydrogen to impinge upon a few cubic centimetres of water contained in a platinum dish kept cold by floating on ice and water. On adding to this a solution of potassium iodide and starch, a blue colour appears, indicating the liberation of iodine, which is more pronounced in the presence of a few drops of dilute sulphuric acid. A higher oxide of hydrogen, the peroxide, has been produced in minute quantity. This method, with modifications, has actually been applied to the manufacture of hydrogen peroxide.

A readier method of preparation is found in the action of dilute sulphuric acid upon barium peroxide, previously ground to a fine powder and made into a thin paste with water. If the temperature be not allowed to rise above 10° C. or so, the reaction proceeds thus:

$$BaO_2 + H_2SO_4 = BaSO_4 + H_2O_2.$$

Since the barium sulphate formed is insoluble in water, the particles of barium peroxide rapidly become coated with a protecting film of barium sulphate and the action ceases. It is therefore expedient to continue the grinding during the reaction so as to break up the solid particles and expose fresh surfaces to the action of the acid. An excess of barium peroxide is used, and the grinding continued until the re-action is only faintly acid. On filtering, a dilute solution of hydrogen peroxide is obtained containing a little sulphuric acid as impurity. This may be removed by the cautious addition of barium hydroxide solution and again filtering.

Hydrochloric, phosphoric, hydrofluoric and other acids may be substituted for sulphuric acid in this reaction, or carbon dioxide may be passed into distilled water to which is added very gradually pure barium peroxide made into a cream with water. In all these cases it is usual to evaporate the filtered product, first in air at about 75° C. up to 20 per cent. strength, and then *in vacuo* over concentrated sulphuric acid to about 50 per cent. strength, after which it is extracted with ether, in which hydrogen peroxide is very soluble. The ether is then distilled off from the extract under low pressure.

Although hydrogen peroxide is so unstable, it is possible to distil it at about 85° C. in a current of air without undue loss. Pure hydrogen peroxide is very seldom prepared, and has no technical application. It has however been obtained in colourless anhydrous crystals by cooling a highly concentrated aqueous solution to a low temperature.

An (impure) solution of hydrogen peroxide of moderate concentration, suitable for shewing many experiments, can be made by adding sodium peroxide, Na_2O_2, to water:

$$Na_2O_2 + 2H_2O = 2NaOH + H_2O_2.$$

The solution is however very unstable, owing to the presence of the sodium hydroxide.

The concentration of aqueous solutions is usually expressed in terms of the volume of oxygen obtainable by decomposition from unit volume of the solution. Thus 'perhydrol,' the most concentrated commercial product, which yields 100 times its volume of oxygen, is '100 volume H_2O_2.' The solutions most usually employed for medical use and in the laboratory are 5 volume, 10 volume, and 20 volume H_2O_2.

Quite recently a crystalline compound of urea and hydrogen peroxide of the formula $CO(NH_2)_2 . H_2O_2$, called *hyperol*, has been put on the market. It contains 35 per cent. of H_2O_2 and is quite free from all the usual impurities found in commercial peroxide. The preparation is said to be stable at all ordinary atmospheric temperatures and, when dissolved in ten times its weight of water, produces what is practically 10 volume H_2O_2.

Hydrogen peroxide is unstable under all conditions, though, as has already been indicated, the colder it is kept the slower the rate at which it decomposes into water and oxygen, other conditions being the same.

$$2H_2O_2 = 2H_2O + O_2 + 46200 \text{ cal.}$$

Even this reaction is reversible to some extent. The impinging of superheated steam upon a hot surface and rapid condensation of the product is the essence of at least one patented process for manufacturing hydrogen peroxide.

The decomposition of hydrogen peroxide into water and oxygen is considerably accelerated by catalysts. Finely divided metals such as silver, gold, and platinum; manganese dioxide, colloidal solutions of metals, and many other substances behave thus. On the other hand, traces of phosphoric or sulphuric acid act as 'negative' or retarding catalysts. Perhydrol is usually sold in bottles of glass coated on the inside with a film of paraffin wax, since the silicates of the glass are said to act as positive catalysts, hastening decomposition. It is not claimed that the paraffin is a negative catalyst.

The readiness of hydrogen peroxide to undergo this exothermic decomposition renders it a valuable oxidizing agent, since it parts with oxygen so easily to substances which can take up oxygen.

The term *oxidation* has acquired a wider meaning than simple combination with oxygen. While not capable of being accurately defined, it is a very convenient and widely used expression, and may be taken to include all processes which add on to a substance either oxygen or any element (or group of elements acting as a whole) similar to oxygen; or, processes which remove hydrogen or any element (or group) similar to hydrogen, from a compound. The converse process is *reduction*.

Usually (but not always) oxidation and reduction go on simultaneously. If A oxidizes B, B reduces A. Hydrogen reduces copper oxide to metallic copper, and copper oxide oxidizes hydrogen to water. Hydrogen peroxide oxidizes lead sulphide (black) to lead sulphate (white), the peroxide being simultaneously reduced to water:

$$PbS + 4H_2O_2 = PbSO_4 + 4H_2O.$$

Ozone also oxidizes lead sulphide to lead sulphate, the ozone being 'reduced' to oxygen:

$$PbS + 4O_3 = PbSO_4 + 4O_2.$$

The restoration of oil paintings which have become dark by age through slow conversion of the lead pigment into lead sulphide, is often carried out by washing with a dilute solution of hydrogen peroxide. The action can be illustrated by moistening a piece of filter paper with lead acetate solution, exposing it to the blackening action of a jet of hydrogen sulphide, and then pouring over it a little hydrogen peroxide solution, which bleaches it at once.

The action of hydrogen peroxide upon many substances which are themselves powerful oxidizing agents is peculiar. *Both* are reduced, and oxygen gas is evolved. If a little peroxide be poured on to silver oxide the evolution of oxygen is vigorous, metallic silver in a fine state of division remaining.

$$Ag_2O + H_2O_2 = 2Ag + H_2O + O_2.$$

If excess of hydrogen peroxide be used the evolution of gas slackens but does not cease when all the silver oxide has been reduced, since the metallic silver acts catalytically upon the remaining peroxide.

A similar reaction takes place with potassium permanganate. A dilute solution of this compound in presence of a little sulphuric acid easily parts with oxygen to oxidizable substances. In presence of hydrogen peroxide, oxygen (gas) is evolved, the equation for the reaction being:

$$2KMnO_4 + 3H_2SO_4 + 5H_2O_2 = K_2SO_4 + 2MnSO_4 + 8H_2O + 5O_2.$$

It will be noticed that the oxygen comes from the hydrogen peroxide and from the permanganate in equal proportions. This reaction is sometimes used as a method of estimating the concentration of a solution of hydrogen peroxide, since the violet colour of the permanganate is discharged, the products of the reaction being colourless.

The strong oxidizing agent, potassium dichromate, also reacts in this way. A dilute solution, in presence of a little sulphuric acid, yields oxygen both to oxidizable substances and to hydrogen peroxide:

$$K_2Cr_2O_7 + 4H_2SO_4 + 3H_2O_2 =$$
$$K_2SO_4 + Cr_2(SO_4)_3 + 7H_2O + 3O_2.$$

As in the case of permanganate, one-half of the evolved gas comes from each of these reagents, the dichromate and the peroxide.

In this reaction an intermediate compound is formed, especially in cold solutions, which can be used as a test both for chromates and for hydrogen peroxide. This is the so-called *perchromic acid*, a very unstable substance of a deep blue colour, possibly an additive product of chromic acid and hydrogen peroxide. Since it decomposes rapidly, examination of it with a view to fixing its molecular composition is difficult. The test is applied by adding a solution of hydrogen peroxide to a dilute solution of a chromate or dichromate which has been acidified with dilute sulphuric acid and covered with a layer of ether a few centimetres deep. The perchromic acid dissolves in the ether, giving the latter an intense colour which lasts for some time.

Hydrogen peroxide oxidizes hydriodic acid quantitatively, with liberation of iodine. The reaction is somewhat slow and proceeds thus:

$$2HI + H_2O_2 = 2H_2O + I_2.$$

Indeed, hydriodic acid is so easily oxidized that it is used as a test for oxidizing agents in general, since the liberated iodine is easily seen, especially if a little starch be added to the solution, as has already been indicated.

If the hydriodic acid be present in considerable excess, the rate at which the hydrogen peroxide disappears is proportional to the amount of the latter present at every instant. Since iodine is easily estimated quantitatively, this reaction is frequently utilized for the determination of hydrogen peroxide.

A very sensitive test for traces of hydrogen peroxide is found in a solution of titanic oxide, TiO_2, in dilute sulphuric acid. This solution is nearly colourless, but in presence of a very small quantity of hydrogen peroxide it develops an orange colour, due to the oxidation of TiO_2 to TiO_3. This reaction can, of course, be used in a converse sense as a test for titanium.

Since 34 grm. of hydrogen peroxide on decomposition yield 16 grm. of oxygen, the remainder being water only, the empirical formula must be $(HO)_x$. That x is equal to 2, at all events in aqueous solution, is shewn by a molecular weight determination made by measuring the depression of the freezing point of dilute aqueous solutions.

Chapter III

THE HALOGENS AND THEIR DERIVATIVES

WHEN concentrated sulphuric acid is poured over common salt an energetic reaction takes place with considerable frothing and evolution of a gas, hydrogen chloride, which gives dense and choking fumes in moist air. The gas is very soluble in water, the concentrated solution being the 'muriatic' acid or 'spirit of salt' well known to earlier chemists. The reaction which takes place without heating is represented by the equation:

$$H_2SO_4 + NaCl = NaHSO_4 + HCl.$$

On strongly heating, a further reaction takes place between the bisulphate and common salt thus:

$$NaHSO_4 + NaCl = Na_2SO_4 + HCl,$$

or, adding the two equations:

$$H_2SO_4 + 2NaCl = Na_2SO_4 + 2HCl.$$

The second reaction requires for completion a rather higher temperature than glass vessels can bear with safety.

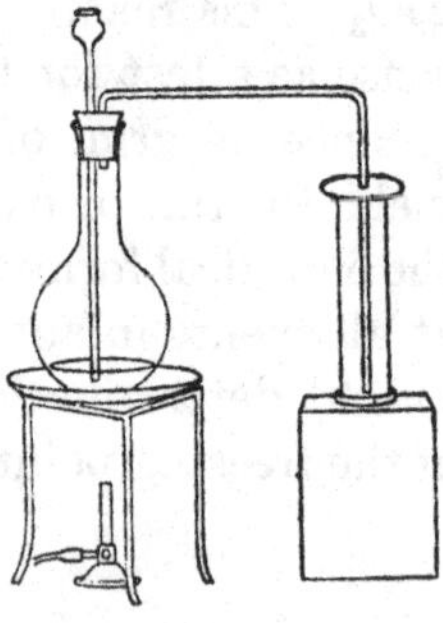

Since the density of the gas is 18·25 in terms of hydrogen as unity, it is somewhat heavier than air and may therefore be collected in an impure condition by displacement of air (Fig. 14), when required to shew qualitative experiments, but when wanted in a purer condition the collection should be made over mercury. As concentrated sulphuric acid has been

Fig. 14

used in its preparation, the gas does not need further drying.

The frothing which occurs when concentrated sulphuric acid is used may sometimes be very inconvenient. If 75 per cent. sulphuric acid (one volume of water with two volumes of concentrated acid) be substituted, there is comparatively little frothing and the stream of gas is steady and continuous. This can be passed through a sulphuric acid wash bottle if required free from moisture.

Hydrogen chloride is not inflammable, nor does it allow other substances to burn in it, but it is nevertheless very reactive. If passed over metals such as sodium, iron, magnesium, aluminium, zinc, etc., contained in a glass tube, and the issuing gas collected, the latter can be shewn to be hydrogen. Hydrogen therefore is a constituent of the gas. The same conclusion is reached by collecting the dry gas in a dry flask, introducing a little mercuric oxide and corking up the flask. Much heat is evolved and moisture condenses in the colder part of the flask. The white solid also formed is mercuric chloride.

That something other than hydrogen is present may be shewn by introducing a little manganese dioxide (which is known to consist of manganese and oxygen only) into a flask of the gas, corking up, and allowing to stand for a few minutes. Again much heat is evolved and the gas acquires a greenish yellow colour. On opening the flask the characteristic odour of chlorine is noticed, and a piece of damp litmus paper introduced into the gas is bleached. Since the chlorine could not have come from the manganese dioxide, it must have been contained in the hydrogen chloride. Chlorine therefore is also a constituent of the gas hydrogen chloride. This is again referred to later.

If a long glass tube be filled with hydrogen chloride and sealed at both ends, on immersing one end in liquid air the gas rapidly condenses, first to a few drops of a water-like liquid, and then to a snow-like solid. In the solid or liquid form it has practically no acid properties. *Water* is necessary for the formation of hydrochloric *acid*.

As has been mentioned, the gas is very soluble. When passed into a flask of ice-cold water until no more dissolves (shewn by the flask and its contents no longer increasing in weight on further passage of gas) the solution has a specific gravity of 1·2 and contains 40 per cent. by weight of hydrogen chloride. The coefficient of absorption in water at 0° C. (and under a pressure of one atmosphere of the gas) is 500; that is, one litre of ice-cold water dissolves 500 litres of hydrogen chloride. In doing so it expands to 1·5 litres. This solution is saturated hydrochloric acid. On exposure to the air it fumes strongly, since the pressure of hydrogen chloride above it is lowered and the gas escapes. If such a solution be subjected to distillation, gas is freely evolved at first and the distillate is saturated with it. The acid in the flask gradually becomes weaker, and the boiling point rises until at 110° C. (if the pressure be 760 mm.) the solution distils without further change; that is, the boiling point remains constant. The distillate at this tempera-ture has the same composition as the acid in the flask, and contains 20·2 per cent. of hydrogen chloride. If the pressure be greater than 760 mm. the boiling point is higher, and consequently more gas is lost. At a pressure of 1500 mm. the 'constant boiling point mixture' contains 19 per cent., and at 50 mm. 23·2 per cent. of hydrogen chloride. Since the composition varies with change of pressure and is not constant over even small ranges, these solutions cannot be regarded as definite chemical com-pounds of hydrogen chloride and water.

If *dilute* hydrochloric acid be distilled, the distillate at first contains very little acid, and the contents of the flask become more and more concentrated until the same constant boiling point concentration mentioned above is attained.

Hydrochloric acid is obtained on a very large scale as a bye-product in more than one industry. In the manufacture of sodium sulphate sulphuric acid is poured upon large masses of common salt, the hydrogen chloride escaping with the furnace gases. Since it is illegal to discharge acid fumes into the air, the hydrogen chloride is removed by means of 'scrubbers.' These are high brickwork towers containing perforated plates placed at intervals and covered with material such as coke, pumice, shingle etc. which is resistent to hydrochloric acid and which offers a large surface for absorption. A stream of water is allowed to percolate down the tower while the furnace gases ascend, passing over the absorbent surfaces. On entering the tower the gases meet with water nearly saturated with hydrogen chloride, while at the top the gases about to escape meet with water free from dissolved gas. The stream of water is so regulated that, while no acid fumes escape into the air, the liquid at the foot of the tower is as nearly as possible saturated. This is the hydrochloric acid of commerce. It always has a pronounced yellow colour due to traces of ferric chloride, which disappears on dilution. The reason for this is given on p. 301.

Sometimes the furnace gases, instead of being 'scrubbed,' are passed through heated vessels containing porous material previously soaked in copper chloride solution. The latter acts catalytically, bringing about a reaction between the hydrogen chloride and oxygen present, thus:

$$4HCl + O_2 = 2Cl_2 + 2H_2O,$$

care being taken that excess of air is present. This is the Deacon process (now obsolescent) for the preparation of chlorine used in the manufacture of bleaching powder. Such chlorine is very impure, being necessarily diluted with nitrogen and other atmospheric gases, but this is to some extent an advantage for the purposes for which it is used, since it prevents the reaction between the chlorine and the slaked lime from proceeding too rapidly and injuring the product by reason of the heat developed* (see p. 97).

It has already been mentioned that hydrogen chloride in an anhydrous condition does not possess acid properties, and therefore the term *hydrochloric acid* is only applied to the solution, which is strongly acid. The reason for this will be seen more clearly later.

The acid reacts readily with many metals, yielding hydrogen and a chloride of the metal. If the latter is capable of forming more than one chloride, as iron ($FeCl_2$ and $FeCl_3$), tin ($SnCl_2$ and $SnCl_4$) and many other metals do, the lower chloride is the one always obtained; that is, the one with the smaller proportion of chlorine. Whenever hydrochloric acid acts upon a metal the products are hydrogen and the '*-ous*' chloride of the metal. Iron gives ferr*ous* chloride, $FeCl_2$; tin gives stann*ous* chloride, $SnCl_2$. On the other hand, when the gas chlorine combines with a metal the higher chloride is formed. Iron gives ferr*ic* chloride, $FeCl_3$; chromium gives chrom*ic* chloride, $CrCl_3$; tin gives stann*ic* chloride, $SnCl_4$. Since there is but one chloride of sodium or of aluminium, it is immaterial whether hydrochloric acid or chlorine be used; the product is the same.

Sometimes a metallic chloride may be obtained in pure condition by evaporating the aqueous solution to crystal-lization or to dryness, as in the case of sodium chloride or

* This mixture of air and chlorine is too dilute to give the best results. Usually it is somewhat enriched by the addition of electrolytic chlorine.

barium chloride, but this is not general. More frequently the solution on evaporation loses acid as well as water, the salt hydrolysing thus:

$$MgCl_2 + H_2O = MgO + 2HCl,$$

either altogether or in part, and, although this is a reversible reaction, the loss of acid results in the final product being either the oxide, or a mixture of the oxide and chloride, or a compound of the two—the *basic* chloride, or oxychloride.

Anhydrous metallic chlorides, if sufficiently volatile, are usually prepared by passing a stream of hydrogen chloride (for the '-ous' chloride) or of chlorine (for the '-ic') over the heated metal, the chloride condensing in the colder part of the apparatus. It is important to remember that the pure chlorides of such metals as iron, chromium, aluminium and many others cannot be obtained from their aqueous solutions by evaporation.

Some metals such as gold and platinum are unattacked by concentrated hydrochloric acid, while others such as copper, mercury, and lead are acted on very slowly. If a small proportion of concentrated nitric acid be added to concentrated hydrochloric acid the mixture, known as *aqua regia*, attacks *all* metals, with the formation of the higher (or '-ic') chlorides, no nitrate being produced. Silver and lead, although readily attacked, form very sparingly soluble chlorides which appear as white and curd-like precipitates.

The nitric acid plays the part of a powerful oxidizing agent, reacting with the hydrochloric acid to form nitrosyl chloride, NOCl, chlorine, and water, thus:

$$HNO_3 + 3HCl = NOCl + Cl_2 + 2H_2O.$$

Aqua regia is undoubtedly more energetic in its action upon metals than moist chlorine. This is probably due to the great chemical energy associated with the production of chlorine in the reaction between nitric and hydrochloric acids.

Hydrochloric acid is readily oxidized, the products being usually chlorine and water. This reaction is utilized for the preparation of chlorine, the oxidizing agents being generally manganese dioxide, potassium dichromate, or potassium permanganate. The reaction between manganese dioxide and concentrated hydrochloric acid was long used as a manufacturing process and is a very convenient one for laboratory preparation. The manganese dioxide and hydrochloric acid are warmed together, the reaction proceeding thus:

$$MnO_2 + 4HCl = MnCl_2 + 2H_2O + Cl_2.$$

The chlorine may be collected over hot water or brine or, in an impure condition, by displacement of air. If necessary, traces of hydrogen chloride may be removed by washing with a little water, but this is of course unnecessary when the gas is collected over hot water or brine. Chlorine is the one important gas which cannot be collected by displacement of mercury.

Pure and dry chlorine may be easily obtained by allowing liquid chlorine to evaporate into vacuous vessels.

The large scale manufacture of chlorine is now almost entirely confined to the electrolysis of brine in the production of caustic soda (see p. 295). When not immediately used it is liquefied and stored in steel cylinders.

Chlorine is a heavy suffocating gas of a greenish yellow colour. It combines readily with most other elements. Phosphorus inflames in the gas, producing the pentachloride, PCl_5, if the chlorine be in excess, or the trichloride, PCl_3, if the phosphorus be in excess. A jet of burning hydrogen introduced into chlorine continues to burn with great energy, forming hydrogen chloride. Hydrogen and chlorine when mixed combine slowly in darkness, but with explosive violence on exposure to a bright light or on the application of a spark. Most metals combine readily with chlorine, and when in the form of very thin leaf or fine powder the action is energetic; powdered antimony, for instance, on being scattered into the gas, burns vividly. Even gold is dissolved slowly by chlorine in aqueous solution.

A candle burns in chlorine with a dull red and very smoky flame, the hydrogen of the candle combining with it to form hydrogen chloride and the carbon being liberated. Certain hydrocarbons, compounds of carbon and hydrogen only, behave similarly. A piece of filter paper moistened with turpentine, $C_{10}H_{16}$, inflames spontaneously when introduced into the gas, producing a dense black smoke.

Perhaps the best known property of chlorine is its power of bleaching many organic colouring substances. Turkey-red cloth and litmus paper, especially if moisture be present, rapidly lose their colour. Writing ink is turned to a pale brown colour, but printer's ink and pencil marks are unchanged, since carbon is not affected. Animal fabrics such as wool and silk cannot be bleached by chlorine, though the latter is in some form or other used almost exclusively for cotton goods. The fibres of wool and silk contain fatty material which is blackened by the gas, this being due to the liberation of carbon. Chlorine is rather soluble in cold water, one volume of water dissolving two and a half volumes of the gas at room temperature, and one volume at 60° C. (the pressure of the chlorine being one atmosphere). When passed into ice-cold water the gas produces a pale yellow crystalline hydrate, to which the formula $Cl_2.8H_2O$ has been assigned; a compound which may be regarded as chlorine crystallized with eight molecules of water.

If chlorine hydrate be packed as tightly as possible in a

glass tube which is then sealed up, on warming the hydrate dissociates, the pressure being sufficient to liquefy the gas, and the latter appears under the water as a heavy yellowish oil. Liquid chlorine was first obtained (accidentally) by Faraday in this way. Chlorine reacts with water slowly in darkness, more rapidly in sunlight, with liberation of oxygen, probably in accordance with the equation:

$$2Cl_2 + 2H_2O = 4HCl + O_2,$$

but this is far from being the whole truth, since the volume of oxygen set free is much smaller than the equation indicates. This can be shewn experimentally by filling a U-shaped eudiometer with chlorine water, displacing some of the liquid in the open limb and then corking. On standing in sunlight, oxygen collects in the closed limb. The liquid on being transferred to an open dish and boiled, or allowed to stand until the odour of chlorine has disappeared, shews an acid reaction.

The reversibility of this reaction is better understood by considering the thermal changes involved. The reaction between hydrogen chloride and oxygen is exothermic, as can be seen from the heats of formation of the various compounds concerned.

Heats of formation per gramme-molecule, ($K = 1000$ cal.)

H_2O (steam), $58K$.

H_2O (liquid), $68 \cdot 4K$.

HCl (gas), $22K$.

HCl, heat of solution of, at infinite dilution*, $17 \cdot 3K$.

In the preparation of chlorine by Deacon's process a temperature of about 430° C. is maintained, when nearly 80 per cent. of the hydrogen chloride is oxidized to chlorine.

---

* Since the heat of solution varies with the concentration, it is usually measured when at its maximum value; i.e., when further dilution gives no additional evolution of heat.

Since everything is in the gaseous condition, the thermo-chemical equation becomes:

$$4HCl + O_2 = 2H_2O + Cl_2 + E.$$
$$4 \times 22K \qquad\qquad 2 \times 58K$$
$$E = (2 \times 58) - (4 \times 22)K.$$
$$E = 28K.$$

When chlorine is hydrolysed by a large quantity of water another factor operates, namely, the heat of solution of hydrogen chloride, the thermochemical equation in this case being:

$$2H_2O + Cl_2 = 4HCl.Aq. + O_2 + E'$$
$$(2 \times 68\cdot4) \qquad\qquad 4 \times 22 + (4 \times 17\cdot3)$$
$$E' = (4 \times 22) + (4 \times 17\cdot3) - (2 \times 68\cdot4).$$
$$E' = 20\cdot4K.$$

Both reactions therefore are exothermic *under the conditions described*. The amount of heat evolved during the formation of a given compound is the same, whether the compound be formed directly or in a series of intermediate stages. This is the law of Hess (1840) and is only a particular case of the general law of the conservation of energy.

Chlorine reacts energetically with caustic alkalis. The nature of the reaction between chlorine and sodium hydroxide in aqueous solution depends upon the temperature, and is typical of the general behaviour of halogens with soluble metallic hydroxides.

When the gas is passed into a cold dilute solution of sodium hydroxide it is absorbed, the products of the reaction being sodium chloride, sodium hypochlorite, and water, according to the equation:

$$Cl_2 + 2NaOH = NaCl + NaOCl + H_2O.$$

An analogous reaction takes place in chlorine water,

$$Cl_2 + HOH = HOCl + HCl.$$

Hypochlorous acid is always present in chlorine water, but it gradually decomposes into hydrochloric acid and oxygen. It will be seen from this that the behaviour of chlorine towards water is not merely a case of simple solution.

The stability of sodium hypochlorite diminishes rapidly with rise of temperature, sodium chlorate and sodium chloride being produced:

$$3NaOCl = 2NaCl + NaClO_3.$$

The products of the action of chlorine on a hot solution of sodium hydroxide are the chloride and chlorate:

$$3Cl_2 + 6NaOH = NaClO_3 + 5NaCl + 3H_2O.$$

This was formerly the usual method of manufacturing potassium chlorate, but, as will be seen from the equation, only one-sixth of the potash and chlorine are converted into the chlorate. The method is now entirely superseded by electrolytic processes. The laboratory preparation of potassium chlorate is simple and easy. If a moderately concentrated solution of potassium hydroxide be saturated with chlorine whilst boiling, a quantity of crystals of potassium chlorate is deposited on cooling. These crystals can be purified from chloride by dissolving in hot water, cooling, draining off the product with the aid of a filter pump, and finally washing with small quantities of cold water. Since the chlorate, although very soluble in hot water, dissolves but sparingly in cold, while the solubility of the chloride in cold water is considerable, these operations can be carried out with but little loss.

The preparation of bleaching powder is an application of the former of these reactions. Chlorine is admitted into chambers filled with trays containing slaked lime, $Ca(OH)_2$, so long as there is any absorption of the gas, the temperature being kept low to avoid the formation of chlorate. The interaction between the chlorine and the slaked lime resembles, *mutatis mutandis*, that between chlorine and soda or potash in *cold* solution:

$$2Cl_2 + 2Ca(OH)_2 = CaCl_2 + Ca(OCl)_2 + 2H_2O.$$

This however cannot be the whole truth. In the first place, according to the equation, slaked lime should absorb approximately its own weight of chlorine. In practice, it cannot be made to take up more than about half this quantity. Secondly, when bleaching powder is shaken with water a solution is obtained which gives the reactions of hypochlorites, but an undissolved residue remains which is chiefly slaked lime.

It is highly improbable that the latter can be present as calcium hydroxide in the bleaching powder, since it no longer absorbs chlorine. The soluble portion again cannot be regarded as a simple mixture of calcium chloride and calcium hypochlorite, since the former is an extremely deliquescent salt whereas bleaching powder shews no tendency towards deliquescence. Moreover calcium chloride dissolves readily in alcohol, but calcium chloride cannot be obtained from bleaching powder by extracting it with this solvent. Odling considered the active constituent of this material, i.e. the soluble portion, to be a compound of chloride and hypochlorite having the formula $CaCl_2.Ca(OCl)_2$, or more simply, $CaOCl_2$. Such a formula ignores the insoluble portion. Other formulae have been proposed which include the basic constituent, but these have only a qualitative significance. It is now fairly well recognized that bleaching powder is not a *single* chemical individual, and therefore there is no justification for assigning to it any chemical formula.

Bleaching powder is much used in calico bleaching and printing; as a disinfectant, and also as a very convenient source of small quantities of chlorine. It has a pronounced odour of hypochlorous acid, which is liberated from it by the action of all dilute acids, including atmospheric carbon dioxide in the presence of moisture. Hypochlorous acid in presence of hydrochloric acid yields chlorine thus:

$$HOCl + HCl = H_2O + Cl_2.$$

The basis of the technical operation of bleaching is a preliminary soaking of the fabric in a dilute solution of bleaching powder, and a subsequent treatment with dilute sulphuric acid which liberates chlorine.

Molecular Formula of Hydrogen Chloride. If equal volumes of hydrogen and chlorine be mixed and exposed to bright light, combination takes place without change of volume. An experiment to prove this may be carried out by filling thick-walled glass tubes of about 25 cm. length and 1 cm. internal diameter, fitted at both ends with capillary stopcocks, with a mixture of equal volumes of hydrogen and chlorine. This is best done by connecting the tubes in series to an apparatus for delivering gas obtained by the electrolysis of concentrated hydrochloric acid previously saturated with chlorine. The gas is dried by passing through concentrated sulphuric acid before admission to the tubes, the operation being carried out in darkness. It is found that hydrogen and chlorine are obtained in equal volumes after electrolysis has proceeded for about one hour, and at the end of this time one of the tubes is tested by opening it under an aqueous solution of potassium iodide. If the mixture consists of equal volumes of the two gases the liquid will half fill the tube, since the chlorine is absorbed with production of a brown solution of iodine in potassium iodide, the hydrogen of course remaining undissolved. When equality of the volumes is thus assured, a second tube is exposed to the light of burning magnesium ribbon, which brings about the union of the gases with slight explosion. After a brief interval to allow the contents of the tube to regain the temperature of the room, the end of the tube is immersed in mercury and the stopcock opened. No gas escapes, nor does the mercury enter the tube, thus shewing that the volume of the gas is unchanged, since the tubes were originally filled at atmospheric pressure. The tube is again opened under potassium iodide solution when the liquid completely fills it, thus shewing the absence of uncombined hydrogen. The absence of uncombined chlorine is indicated by the liquid remaining colourless.

Since one volume of hydrogen combines with one volume of chlorine to produce two volumes of hydrogen chloride, it follows from the application of Avogadro's hypothesis that one molecule of hydrogen combines with one molecule of chlorine to produce two molecules of hydrogen chloride. Obviously the molecules of hydrogen and of chlorine have been divided, and therefore each must have consisted of *not less* than two atoms. The evidence that *not more* than two atoms are present in each molecule is obtained from a study of the reactions of the acid itself. No attempt to remove either the hydrogen or the chlorine from it *in fractions* has ever succeeded; either the whole of the hydrogen (or chlorine) is displaced from the molecule, or none. This is of course not *proof* that only one atom of each of the elements chlorine and hydrogen is present in the molecule of hydrogen chloride; it only indicates probability. In view however of accumulated experience of the reactions of this compound, it has become probability of so high an order that the diatomicity of the molecules of hydrogen and chlorine is universally accepted; a view which is supported by further evidence which will be found in works on physical chemistry. The reactions described above can therefore be represented by the molecular equation:

$$H_2 + Cl_2 = 2HCl.$$

Fluorine. Chlorides and fluorides resemble each other in their reactions with sulphuric acid. When the concentrated acid is warmed with powdered fluor spar (calcium fluoride, CaF_2) hydrogen fluoride is evolved.

$$CaF_2 + H_2SO_4 = CaSO_4 + 2HF.$$

The operation is carried out in leaden vessels, since hydrogen fluoride attacks glass and other siliceous materials (see Chap. IX). The gas dissolves freely in water forming hydrofluoric acid, which must be kept in vessels of lead or gutta percha. It is used technically for etching glass.

Hydrofluoric acid is much more stable than hydrochloric acid. There is a marked difference between the two in their behaviour towards oxidizing agents. Chlorine is readily obtainable by the oxidation of hydrochloric acid, but fluorine cannot be liberated from its acid by any process of oxidation.

Since the element is extremely reactive its preparation is attended by great experimental difficulties, as the containing vessels themselves are attacked. Fluorine was first isolated by Moissan (1886) by the electrolysis of a mixture of hydrogen fluoride and potassium fluoride in a platinum apparatus with platinum-iridium electrodes (Fig. 15). It is an almost colourless gas of a very penetrating odour, combining with hydrogen explosively even in darkness and at very low temperatures. It decomposes water rapidly, forming hydrofluoric acid and liberating oxygen, partly in the form of ozone. This element is remarkable in so far as it does not combine with oxygen under any conditions, notwithstanding its pronounced reactivity.

Bromine and Iodine. Bromides and iodides react with concentrated sulphuric acid similarly to fluorides and chlorides, the halogen hydride being produced:

$$KBr + H_2SO_4 = KHSO_4 + HBr,$$

$$KI + H_2SO_4 = KHSO_4 + HI.$$

In neither case however is the pure gas produced, since both compounds are oxidized by concentrated sulphuric

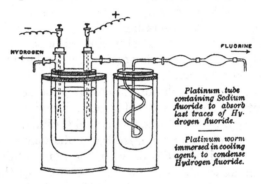

Platinum tube containing Sodium fluoride to absorb last traces of Hydrogen fluoride.

Platinum worm immersed in cooling agent, to condense Hydrogen fluoride.

Fig. 15

acid with liberation of the halogen. When hydrogen bromide reduces sulphuric acid the products of the reaction are sulphur dioxide, bromine, and water:

$$2HBr + H_2SO_4 = Br_2 + 2H_2O + SO_2.$$

Hydrogen iodide being more unstable than hydrogen bromide has a more energetic action, the sulphuric acid being reduced to sulphur dioxide, free sulphur, or hydrogen sulphide, according to the experimental conditions. The changes may be represented in the following stages:

$$H_2SO_4 + 2HI = 2H_2O + SO_2 + I_2,$$

$$SO_2 + 4HI = 2H_2O + S + 2I_2,$$

$$S + 2HI = H_2S + I_2.$$

The reaction may however proceed by direct reduction of sulphuric acid to sulphur or hydrogen sulphide thus:

$$H_2SO_4 + 6HI = 4H_2O + S + 3I_2,$$
$$H_2SO_4 + 8HI = 4H_2O + H_2S + 4I_2.$$

These products, namely iodine, sulphur dioxide, sulphur, and hydrogen sulphide, together with unchanged hydrogen iodide, may all be recognized by covering a few crystals of potassium iodide in a test tube with concentrated sulphuric acid and gently warming.

If sulphuric acid considerably diluted with water be employed, no free halogen is obtained nor is the sulphuric acid reduced. For analytical purposes hydriodic acid is used in the form of a solution of potassium iodide acidified with dilute sulphuric acid. Hydrogen bromide may be prepared by the action of sulphuric acid, diluted with an equal volume of water, on crystals of potassium bromide. The gas is freely evolved on warming and contains but little bromine, which may be removed by passage through a U-tube containing small pieces of moist phosphorus, since phosphorus reacts readily with bromine to form the tri-bromide, PBr_3, and the latter is hydrolysed by water forming phosphorous acid and hydrobromic acid. These reactions proceed simultaneously:

$$PBr_3 + 3H_2O = H_3PO_3 + 3HBr.$$

The gas may be dried by calcium bromide. Obviously concentrated sulphuric acid cannot be used for this purpose.

This method is not applicable to the preparation of hydrogen iodide, which is generally obtained by the interaction of iodine, phosphorus, and water. The tri-iodide of phosphorus hydrolyses in the same way as the tri-bromide:

$$PI_3 + 3H_2O = H_3PO_3 + 3HI.$$

In practice it is not convenient to use the tri-iodide directly as the operation is somewhat dangerous. Iodine and red phosphorus are placed together in a flask (Fig. 16) and water is slowly dropped upon the mixture. The gas is evolved freely and passed through a U-tube containing phosphorus precisely as in the case of hydrogen bromide. If excess of iodine be present, some phosphoric acid is also produced:

$$2P + 5I_2 + 8H_2O = 2H_3PO_4 + 10HI.$$

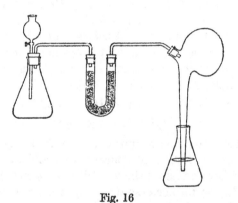

Fig. 16

The method may also be employed for the preparation of hydrogen bromide, but in this case the water and phosphorus are placed together in the flask and bromine is slowly dropped on the mixture from the funnel.

These examples may be regarded as special cases of a general reaction. Substituted waters, or alcohols; that is, compounds in which one of the hydrogen atoms of the water molecule is replaced by a group of elements functioning as a whole, behave in an analogous way. For example, methyl alcohol, CH_3OH, and ethyl alcohol, C_2H_5OH, react with phosphorus together with bromine or iodine to form the corresponding methyl and ethyl halides; methyl alcohol yields methyl bromide, CH_3Br, or methyl iodide, CH_3I; ethyl alcohol the parallel ethyl halides, C_2H_5Br and C_2H_5I. In such cases the apparatus employed requires modification when the product is a liquid at the ordinary temperature.

The halogen hydrides are very soluble in water. Such solutions all resemble hydrochloric acid, the concentrated solutions becoming more dilute and the very dilute acids more concentrated on distillation, until the constant boiling point mixture of each is attained.

A moderately concentrated solution of hydrobromic or hydriodic acid can be made by passing sulphur dioxide or hydrogen sulphide through water in which the corresponding halogen is suspended or dissolved:

$$SO_2 + 2H_2O + Br_2 = H_2SO_4 + 2HBr,$$
$$H_2S + I_2 = 2HI + S.$$

In the latter case the precipitated sulphur can be removed by filtration, and the acid concentrated and at the same time freed from excess of the volatile hydrogen sulphide by boiling. In the former case distillation is necessary if the pure acid be required.

Bromine and iodine generally occur together in nature. The former is almost exclusively obtained from certain lacustrine deposits, notably those of Stassfurt, in which it is found in the form of bromides of certain metals, chiefly potassium and magnesium. The mother liquors from the separation of potassium chloride and carnallite ($KCl \cdot MgCl_2 \cdot 6H_2O$) are rich in bromides. When chlorine is passed into this liquor bromine is liberated and the corresponding chloride produced

$$2KBr + Cl_2 = 2KCl + Br_2.$$

The bromine is distilled off and condensed as a dark red

volatile liquid of a penetrating disagreeable odour. A similar method is now in operation in America for recovering bromine from sea-water.

The laboratory preparation of bromine is analogous to the preparation of chlorine from common salt. Potassium bromide and manganese dioxide are mixed together and warmed with moderately concentrated sulphuric acid, when bromine distils over:

$$2KBr + MnO_2 + 3H_2SO_4 = 2KHSO_4 + MnSO_4 + 2H_2O + Br_2.$$

Bromine and mercury are the only elements which are liquid at ordinary laboratory temperatures. The former has a specific gravity of 3·18 at 0° C., boils at 63° C., and freezes at − 7° C. Water dissolves 4·17 per cent. of its weight of bromine at 0° C., and 3·6 per cent. at 20° C. It forms a hydrate of the formula $Br_2 . 10H_2O$. Bromine vapour is most destructive to the mucous membrane, while the liquid when it comes in contact with the skin produces painful wounds which heal with difficulty. In many respects, notably in its behaviour towards solutions of caustic alkalis, it closely resembles chlorine.

Iodine is now obtained chiefly from Chile saltpetre (caliche) which is mainly sodium nitrate, and which is found in vast quantities in the rainless districts of South America. This material on purification by crystallization leaves large quantities of sodium iodate in the mother liquors, from which iodine is obtained by reduction with a mixture of sodium sulphite and bisulphite:

$$2NaIO_3 + 3Na_2SO_3 + 2NaHSO_3 = 5Na_2SO_4 + H_2O + I_2,$$

the iodine being filtered off, dried, and purified by sublimation.

The old method of obtaining iodine from seaweed is now nearly obsolete, but is occasionally revived in the British Isles, Brittany, and elsewhere. The ashes of burnt seaweed (chiefly *Laminaria digitata*) are extracted with water, when the solution contains small quantities of sodium iodide and sulphide. This liquid is acidified with dilute sulphuric acid, when hydrogen sulphide and sulphur are liberated and removed. The dilute hydriodic acid remaining is oxidized with potassium bichromate, the liberated iodine being then allowed to settle. It is subsequently filtered off, dried, and sublimed.

The laboratory preparation of iodine from potassium iodide is in all respects similar to that of bromine from potassium bromide, the apparatus being slightly modified as iodine is a volatile solid.

The element is a brownish black solid with a high lustre. On gently warming it vaporizes without melting, since under atmospheric pressure its boiling point is lower than its melting point; or in other words the vapour pressure of solid iodine is greater than that of liquid iodine. If the pressure of the vapour be increased, iodine melts to a brown liquid. The gas, which has a fine violet colour, condenses on a cold surface in crystals.

The vapour density varies with change of temperature, as also does the colour. Up to about 500° C. the vapour density is 127, corresponding to a molecular weight of 254 and a diatomic molecule, I_2. Above 500° C. the vapour density falls, gradually approaching a constant value of 63·5, indicating a monatomic molecule, I, at 1500° C. Above this latter temperature the density remains constant, as would be expected.

$$I_2 \rightleftarrows 2I \mp 35700 \text{ cal. (Bjerrum).}$$

Iodine is very sparingly soluble in water. Certain organic solvents such as chloroform and carbon disulphide dissolve it freely, yielding violet solutions, while the solution in alcohol or ether has a brown colour. These differences in colour are attributed to differences in molecular aggregation.

When a solute is shaken with two non-miscible solvents until equilibrium is attained, the solute distributes itself between the two solvents in a ratio which is constant for a given temperature, and which is independent of the mass of the two solvents. This is known as the *distribution ratio* or *partition coefficient*, and is equal to the ratio of the solubilities of the solute in those solvents.

If c_1 denotes the concentration in solvent A and c_2 that in solvent B, both expressed in grammes of solute per cubic centimetre of solution, the solute will always distribute itself between the solvents in these proportions, independently of the mass of each solvent present, and the ratio c_1/c_2 is constant for varying states of dilution at a particular temperature. This constancy only holds when the molecular weight of the solute is identical in both solvents. Iodine is 131 times as soluble in chloroform as in water, at laboratory temperature. If iodine be shaken with a mixture of chloroform and water in any proportions until equilibrium is attained, and the two liquids allowed to separate, the mass of iodine per cubic centimetre of the chloroform solution will always be 131 times the mass of iodine per cubic centimetre of the aqueous solution, whatever the total mass of iodine used. This principle is much used qualitatively in analytical operations and in organic preparations. Iodine and bromine are identified by extracting them from their aqueous solutions with chloroform, carbon disulphide, or ether. The extraction of silver from lead by means of zinc depends upon the same principle. The silver-containing lead is melted with zinc and the molten mixture well agitated. On standing, the molten zinc floats to the surface bringing with it in solution most of the silver (see Parkes's process, Chap. XI).

Unlike chlorine, which is less soluble in solutions of metallic chlorides than in water, iodine dissolves freely in aqueous solutions of metallic iodides, giving a brown colour. A solution of potassium iodide is generally used as a solvent for iodine, both for analytical and preparative operations. The solubility of iodine increases with the concentration of the potassium iodide. No doubt a compound of potassium iodide and iodine exists in the solution, but this is so easily dissociated that the element reacts in most cases as though it were merely dissolved (see also Chap. x).

Traces of iodine in aqueous solution are easily detected by means of the intense blue colour iodine gives with a solution of starch. The colour disappears on heating but returns when the liquid is cooled.

Iodine reacts with solutions of caustic alkalis in a manner not unlike that of chlorine. With a cold and dilute solution of potash the reaction possibly begins thus:

$$2KOH + I_2 = KOI + KI + H_2O,$$

potassium hypoiodite however, if formed at all, is so unstable that it is at once resolved into iodate and iodide:

$$3KOI = KIO_3 + 2KI.$$

Practically therefore the only products of the reaction which can be isolated are potassium iodide and iodate, whatever the temperature may be:

$$6KOH + 3I_2 = 5KI + KIO_3 + 3H_2O.$$

Formerly this was the usual way of preparing potassium iodide. Iodine was dissolved in potash, powdered charcoal stirred into the solution, the mass evaporated to dryness and ignited. The function of the charcoal was to assist in the reduction of iodate to iodide. The latter was extracted with water from which it crystallized after filtration. Potassium iodide thus made is liable to contain traces of iodate. These may easily be detected by acidifying with dilute hydrochloric or sulphuric acid. The iodide and iodate of potassium do not interact, but on acidifying, the corresponding acids, hydriodic and iodic, are liberated. These cannot co-exist since they react together thus:

$$5HI + HIO_3 = 3H_2O + 3I_2,$$

and the liberated iodine, even in minute quantity, can be identified by the addition of starch solution.

A method of preparing potassium iodide which gives a product free from iodate consists in warming together iron filings, iodine, and water till all the iodine has gone into solution as ferrous iodide. On filtering from the excess of iron and oxidizing the ferrous iodide to the ferric condition by further addition of iodine, a solution containing ferric iodide is obtained. This is boiled with a slight excess of potassium carbonate and filtered. On evaporation of the filtered solution, crystals of potassium iodide are deposited.

Chlorine and iodine combine, forming iodine monochloride, ICl, and the trichloride, ICl_3. The former is obtained when dry chlorine comes in contact with dry iodine as a bromine-like liquid which solidifies on cooling. The compound is known in two modifications, having the same composition but different melting points. The trichloride is formed by the action of excess of chlorine on iodine or on the monochloride. It is an orange-yellow crystalline solid.

The hydrides of the halogens differ remarkably in stability, which diminishes as the atomic weight of the halogen ascends. As has been shewn, hydrogen fluoride is extremely stable and resists oxidation under all conditions. Hydrogen chloride can be oxidized by manganese dioxide and other moderate oxidizing agents. Hydrogen bromide and iodide are easily oxidized, as has been seen by their action on concentrated sulphuric acid. Both are dissociated on heating, the iodide more easily than the bromide. Solutions of hydriodic acid rapidly develop a brown colour from the liberation of iodine. This acid, owing to its instability, is a powerful reducing agent. The progressive diminution in stability is further shewn in the heats of formation:

Element	Atomic weight	Heat of formation of the hydride
Fluorine	19	38·6 K
Chlorine	35·5	22·0 K
Bromine	80	8·4 K
Iodine	127	−6·1 K (endothermic)

The stability of the oxygen derivatives *increases* as the atomic weight of the halogen ascends. Fluorine forms no oxygen compound: chlorine and bromine form oxides which are unstable, though the chlorates and bromates are well-defined and stable salts: iodine forms a pentoxide, I_2O_5, corresponding to iodic acid HIO_3. This may be obtained by boiling iodine with nitric acid, filtering off the undissolved iodic acid and gently heating the latter, when the pentoxide remains as a white and fairly stable solid.

The diminishing energy of reaction of the halogens with metals is analogous to that with hydrogen. The halogen of lower atomic weight will always displace one of higher atomic weight from a solution of a halide. Chlorine liberates both bromine and iodine from solutions of bromides and iodides. Bromine in like manner displaces iodine from iodides, but not chlorine from chlorides. If chlorine be passed into a solution containing a bromide and an iodide, it liberates the whole of the iodine before any bromine is set free. These reactions are of importance in chemical analysis.

The halides of silver are very sparingly soluble in water and dilute acids. They all undergo a change on exposure to light, the colourless chloride rapidly darkening to a deep violet. This property is utilized in photography. Sensitive plates, films, and printing paper are made by coating glass, celluloid, or paper with an emulsion of the silver halide in gelatine. On exposure to light the halide, in presence of gelatine, is reduced. Development consists in furthering this process by means of suitable reducing agents, the reduction taking place more rapidly on those points where the process has been started by the action of light. When a sufficiently pronounced picture in finely divided metallic silver has been produced, reduction is stopped by 'fixing,' that is, by dissolving away the remaining silver halide by a solution of sodium thiosulphate, $Na_2S_2O_3$, sometimes erroneously termed hyposulphite. The thiosulphate is removed by washing in water.

The process may be shewn very simply. A piece of filter paper is soaked in dilute silver nitrate, dried, dipped for a few seconds in dilute sodium chloride, and again dried, in darkness. If this paper be covered with a sheet of wire gauze and exposed to light a print of the pattern of the gauze is obtained, which can be rendered permanent by soaking in sodium thiosulphate solution and finally washing. The process of 'toning,' which consists in replacing the metallic silver by some other metal, can be illustrated by placing the washed print in a dilute solution of the chloride of gold or of platinum, when the colour gradually changes as the exchange of metals takes place.

CHAPTER IV

CARBON AND ITS SIMPLER COMPOUNDS

THE element carbon occurs in nature both in the free state and to a much greater extent in combination. In the mineral world it is a constituent of metallic carbonates and of rock oil. Coal is of vegetable origin and may contain as much as 87 per cent. of carbon.

This element is an essential constituent of all living matter and exists in most products of vital activity. In the atmosphere it is present in the form of carbon dioxide to the extent of about three parts in ten thousand. Its compounds are far more numerous than those of all the other elements added together, and for this reason it is usual to divide chemical studies into Organic chemistry, which deals with the compounds of carbon, and Inorganic chemistry, which is concerned with the other elements and their compounds. As a few of the simpler compounds of carbon are of great importance in relation to other elements, this division is not rigid.

The existence of vast numbers of organic compounds is due to the remarkable power of the carbon atom of combining with other carbon atoms. The number of atoms of a particular element, other than carbon, which are found linked together directly in a molecule, is limited and usually quite small. In the case of carbon compounds the number of carbon atoms which can be linked together seems to be limited only by the difficulty of experiment. In the paraffin series of hydrocarbons—a series represented by the general formula C_nH_{2n+2}—a compound containing sixty carbon atoms linked together has been obtained, and there is no indication that still larger molecules cannot exist.

No other element approaches carbon in respect of this power of linkage. Its nearest neighbour, silicon, shews the phenomenon to a slight extent but is not comparable with carbon in this property.

Allotropy of Carbon. Carbon is known in two crystalline modifications, diamond and graphite, and also in an amorphous condition of varying degrees of purity such as coke, soot, gas carbon, charcoal, etc. Diamond is exceedingly hard and usually transparent. True diamonds are identified by their specific gravity, 3·52, by their refractive index, 2·42, and by their resistance to the action of hydrofluoric acid and fused potassium hydroxide. At a red heat they are unchanged in presence of inert gases, but burn in oxygen with the formation of carbon dioxide. At the temperature of the electric arc they are converted into the graphitic modification of carbon.

The great hardness of diamond renders it useful for such purposes as glass cutting and for facing drills for boring hard rocks. Microscopic diamonds have been made by Moissan by suddenly cooling a solution of carbon in molten iron. Such iron expands slightly on passing from the liquid to the solid phase, and since the surface of the molten iron solidifies before the interior, the cooling of the internal mass with the consequent separation of the dissolved carbon takes place under very great pressure. On dissolving away the iron in acid the diamonds remain, but such artificial diamonds are not comparable either in size or appearance with the natural product.

The second form of crystalline carbon, graphite, occurs naturally in various parts of the world, and is also prepared artificially as a bye-product in the manufacture of carborundum. It is a greyish black lustrous solid, having a specific gravity of about 2·1, and is a good conductor of electricity. Graphite has considerable value as a lubricant;

for polishing certain metals, such as lead shot, and iron stoves; for making 'lead' pencils, the 'lead' being a mixture of graphite and china clay in varying proportions according to the hardness required in the pencil; and for graphite crucibles and similar vessels required to withstand high temperatures. This variety of carbon is more easily oxidized than diamond. The latter is not attacked by liquid oxidizing agents, whereas graphite may be oxidized to a curious substance known as graphitic acid by treatment with a mixture of concentrated nitric and sulphuric acids.

Under the general name of amorphous carbon are included a number of substances, consisting essentially of the element, but differing considerably in physical properties and in the amount of impurity present. The various forms of charcoal may be prepared by heating certain organic substances to redness out of contact with air. Wood is commonly used for this purpose but a purer product may be obtained from sugar. For special purposes the charcoal made by heating in a similar manner blood, bones, ivory, or coco-nut shells finds considerable application. Coke, lampblack, and soot are also forms of amorphous carbon. Another kind of charcoal, which approaches more closely to graphite than any of the others, is that obtained from the retorts of gas works and known as gas carbon or retort carbon. It is very dense and hard and is of considerable importance in electrical work, being used for electrodes and the pencils of arc lamps. In purely chemical properties the various kinds of amorphous carbon differ only qualitatively from graphite. They are more easily attacked by oxidizing agents. Wood charcoal is oxidized to carbon dioxide on heating with concentrated sulphuric acid. On prolonged treatment with a solution of potassium permanganate it is oxidized in part to mellitic acid, $C_6(COOH)_6$. This reaction throws considerable light on the molecular structure of charcoal, though the acid itself is of little importance. Since permanganate tends to break down the linking between carbon atoms rather than to cause carbon atoms to unite together, it points to the improbability of the carbon molecule in charcoal containing *less* than twelve atoms.

Certain forms of charcoal have a wonderful power of adsorbing gases, which has been turned to account in producing high vacua. Coco-nut charcoal cooled in liquid air was first used for this purpose by Dewar. In preparative work solutions are frequently freed from colouring impurities by boiling with one or other of the forms of animal charcoal, the process being one of adsorption.

Oxides of Carbon. When any of the modifications of carbon are burnt in air two oxides are formed, the monoxide, CO, and the dioxide, CO_2. It is difficult to determine which oxide is the first to be produced, since the dioxide is easily reduced to the monoxide by red hot carbon. When 12 grm. of carbon (charcoal) are oxidized to the monoxide, $29 \cdot 6 \, K$. are evolved. On further oxidation to dioxide an additional $68 \cdot 0 \, K$. are produced, the union of the second atom of oxygen being therefore attended with a much greater liberation of energy than the first. Since carbon monoxide is used both for heating purposes and for power on an enormous scale, and since the monoxide is regenerated from the dioxide, this thermal relationship is of great importance.

In the combustion of carbon under ordinary conditions the proportion of carbon monoxide formed is not great, but its presence can be observed as it burns with a characteristic blue flame on the surface of a glowing coke fire, where the monoxide meets with a further supply of atmospheric oxygen. Even in a burning cigarette an appreciable quantity of carbon monoxide is formed.

Producer gas consists mainly of carbon monoxide and nitrogen, and is made by passing furnace gases or air over red hot coke.

$$CO_2 + C = 2CO.$$

The impure carbon monoxide thus obtained is used for heating metallurgical furnaces, and for driving gas engines. The diluting effect of the nitrogen is a great disadvantage but it cannot economically be avoided.

Water gas is the product obtained by blowing steam over red hot coke.

$$C + H_2O = CO + H_2 - 28 \cdot 4\,K.$$

As will be seen from the equation, the gas consists mainly of carbon monoxide and hydrogen, but some carbon dioxide and nitrogen are always present. As the reaction is strongly endothermic, the temperature of the coke falls rapidly. At intervals therefore the passage of the steam is interrupted, and the temperature again raised by blowing a current of air over the coke. The formation of water gas is necessarily intermittent, a difficulty surmounted by using several furnaces, so that the supply of gas may be practically uninterrupted, by the simple operation of switching over from one generating furnace to another.

A modified kind of water gas, known as *semi-water gas*, is produced as a continuous process by admitting with the steam sufficient air to render the whole reaction slightly exothermic. In this gas there is a larger proportion of useless nitrogen, but as a fuel it is extraordinarily cheap.

Hydrogen is obtained from water gas on a large scale for technical purposes by removing the carbon monoxide. The latter gas reacts with steam at a bright red heat, or at lower temperatures in presence of a suitable catalyst, such as iron or nickel, to produce carbon dioxide and hydrogen:

$$CO + H_2O = CO_2 + H_2,$$

the carbon dioxide being removed from the hydrogen by absorption in milk of lime.

Although carbon dioxide is the final product of oxidation of every variety of carbon, this does not provide a convenient method of preparation of the gas for experimental

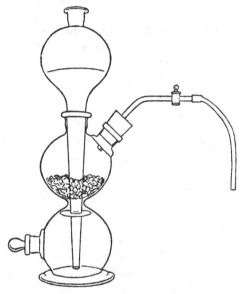

Fig. 17

purposes. It is usually made in the laboratory by acting upon marble with hydrochloric acid, and a supply can be obtained as desired by the use of a Kipp's apparatus (Fig. 17).

$$CaCO_3 + 2HCl = CaCl_2 + H_2O + CO_2.$$

Practically any acid with any carbonate will yield the gas for a time, but if the salt of the acid produced is sparingly soluble in water, the action, which is a surface one, is soon brought to an end by the protective coating of insoluble salt. Sulphuric acid, for instance, cannot be used for the preparation of the gas from marble, owing to the sparing solubility of calcium sulphate. When sodium carbonate is used it is immaterial which acid is taken, since sodium salts are practically all soluble in water.

Many carbonates and all bicarbonates yield carbon dioxide on heating, though the temperatures at which the evolution of the gas takes place vary widely. The bicarbonates without exception decompose at quite low temperatures. The dissociation of sodium bicarbonate according to the equation

$$2NaHCO_3 = Na_2CO_3 + H_2O + CO_2$$

takes place appreciably at about 30° C., either in the solid condition or in solution; indeed a solution of sodium bicarbonate can only be prepared by keeping the solvent cool, and is rapidly and completely converted into the normal carbonate by boiling, and this behaviour is typical of bicarbonates in general. The normal carbonates of most metals when heated dissociate into the oxide of the metal and carbon dioxide. This indeed is one of the general methods of preparing basic oxides, and is widely applicable. Magnesium carbonate dissociates below a red heat; calcium carbonate requires a bright red heat; barium carbonate hardly dissociates appreciably below the melting point of platinum (circa 1750° C.), whereas sodium carbonate is unaffected by any temperature attainable under ordinary laboratory conditions.

If calcium carbonate be heated in a closed vessel provided with a pressure gauge, it will be found that the pressure of the carbon dioxide assumes a definite value which depends solely upon the temperature. Dissociation begins to be perceptible at about 440° C. (according to Debray) and reaches atmospheric pressure at a little above 800° C. From this point the pressure increases rapidly for small rises of temperature, as the following figures, obtained by Le Chatelier, shew:

Temperatures	610°	810°	812°	865° C.
Pressures (mm. mercury)	46	678	753	1333

The pressure of the carbon dioxide, known as the *dissociation pressure*, is in no way dependent upon the relative quantities of calcium carbonate and lime present. Such a condition of equilibrium is in accordance with the phase rule, there being two solid phases and a gas. The system is univariant, the components being lime and carbon dioxide.

The well-known process of lime burning consists essentially in mixing chalk, limestone, or marble, preferably in large pieces, with about half its weight of fuel, usually coke, the operation being carried out in kilns which hold several tons at a charge. The fire when once started is allowed to burn itself out, the larger pieces of quicklime being afterwards separated from the ash and dust. These latter find some application in agriculture.

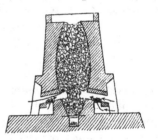

Fig. 18

From the figures quoted above it will be seen that the temperature of the lime kiln should be well above 800° C., otherwise the process is slow. If the temperature be too high, the lime is liable to react with any clay which may be present, forming a silicate of calcium, a substance which renders the quicklime very slow in taking up water. Such lime is said to be 'dead burnt.'

When water is poured upon quicklime much heat is

evolved. The solid at first falls to a powder known as slaked lime or calcium hydroxide. The change is easily reversed by heating.

$$CaO + H_2O \rightleftharpoons Ca(OH)_2.$$

On further addition of water some of the calcium hydroxide dissolves. Milk of lime is a mixture of slaked lime and its saturated solution. Calcium hydroxide is very sparingly soluble, one litre of the saturated solution, that is lime water, containing only 1·7 grm. of the hydroxide at ordinary temperature. In common with that of many calcium salts, the solubility of calcium hydroxide diminishes with rise of temperature.

Builders' mortar is a mixture of milk of lime and sand (impure silica, SiO_2). The *drying* of mortar is merely loss of excess of water, whereas the *setting* is a hardening process brought about by the absorption of carbon dioxide from the air, which results in further loss of water:

$$Ca(OH)_2 + CO_2 = CaCO_3 + H_2O.$$

At this stage the intimate mixture of carbonate, hydroxide and silica is hard, but somewhat friable. In course of time a good mortar further hardens very considerably, in consequence of the slow formation of a silicate of calcium produced by the reaction between the acidic oxide silica, and the basic calcium hydroxide.

Carbon dioxide is also produced during the fermentation of grape sugar (glucose). This can be shewn by adding a little yeast to a solution of glucose, in a flask fitted with a cork and delivery tube which dips into lime water. The main reaction which takes place may be represented by the equation:

$$C_6H_{12}O_6 = 2C_2H_6O + 2CO_2.$$
$$\text{glucose} \qquad \text{alcohol}$$

Since fermentation only takes place between certain limits of temperature, and proceeds most rapidly at about 30° C., the flask should be kept slightly warm.

In the production of 'sparkling' wines and beers, the fermentation is conducted in such a way as to retain the carbon dioxide in solution under pressure. 'Still' wines do not contain carbon dioxide, since fermentation in such cases is carried out in vessels open to the air.

The decay of organic matter is attended by the production of carbon dioxide, and this may account in part for the presence of the gas at the bottom of old disused wells. It is also found issuing from the interior of the earth in certain localities, particularly in volcanic districts. In 1844 Boussingault estimated that carbon dioxide was discharged into the atmosphere at a greater rate by the volcano Cotopaxi than by the city of Paris. The hot and humid atmosphere of the celebrated Grotta del cane contains large quantities of it. In many parts of the world natural springs of water highly charged with the gas occur, as at Vichy, Ems, and Neider-Selters, and in most limestone districts the local springs contain more than the normal amount of carbon dioxide.

Although the addition of carbon dioxide to the atmosphere by the burning of coal and other fuels, by animal respiration, by the decay of vegetation in tropical jungles and forests, by the contribution of volcanic action and in other ways, must be considerable, the proportion of the gas has remained practically constant at three parts in ten thousand during the comparatively short period of time over which measurements have been made. The equation

$$C + O_2 = CO_2$$

represents the loss of oxygen and gain of carbon dioxide by the atmosphere due to carbon combustion of all kinds;

fires, respiration, and so on. On the other hand the equation:

$$nCO_2 + nH_2O = (CH_2O)_n + nO_2$$

is typical of the reaction which carbon dioxide from the atmosphere is made to undergo by the green colouring matter (chlorophyll) of plants, in presence of sunlight. Carbon dioxide is food for the plant and oxygen a waste product. That oxygen is produced may be shewn by fitting up an apparatus as in Fig. 19 and introducing a handful of watercress into the flask. On standing for some time in sunlight, a gas which can be recognized as oxygen collects in the test tube. The carbon of the dioxide is not liberated *qua* carbon, but is built up into carbon compounds, few of which contain so high a proportion of oxygen to carbon as the dioxide itself, and this of course entails a liberation of oxygen. It is not implied that the equilibrium between carbon dioxide and oxygen in the atmosphere is maintained absolutely uniform. Indeed it has been suggested that the relative quantities of these gases have varied enormously at different geological periods.

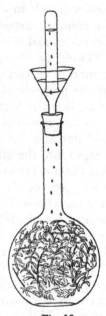

Fig. 19

According to Peligot, the oceans also play a part in regulating the amount of carbon dioxide present in the atmosphere. The reversible reaction between calcium bicarbonate, carbonate, water, and carbon dioxide has already been explained.

$$Ca(HCO_3)_2 \rightleftharpoons CaCO_3 + CO_2 + H_2O$$

Peligot suggests that this reaction occurs in the sea. If the carbon dioxide in the atmosphere increases, the gas tends to go into solution as calcium bicarbonate; if the pressure of carbon dioxide falls, there is a dissociation of the dissolved bicarbonate tending to restore the pressure, and the gas once more finds its way into the atmosphere. Since something like three quarters of the surface of the globe is covered with water, this influence in maintaining the constancy of the carbon dioxide content of the atmosphere is probably considerable, though it does not of course account for the practically constant proportion of oxygen.

Since carbon burns in *excess* of oxygen without producing any change in the volume of the gas* it follows (from Avogadro's hypothesis) that one molecule of oxygen, O_2, produces one molecule of carbon dioxide, and therefore the formula of the latter must be C_xO_2. It has a vapour density of 22 relatively to hydrogen, and hence a molecular weight of 44. Subtracting from this the weight of oxygen present in the molecule, 32, the weight of carbon present is 12, and since this is the atomic weight of carbon, $x = 1$, therefore the molecular formula is CO_2.

The gas is not a supporter of combustion in the ordinary sense. A burning taper plunged into it at once ceases to burn, and some use is made of this property in the direction

* This is shewn as a somewhat elaborate lecture experiment but the student can demonstrate it himself with fair accuracy in the following way:

A small piece of charcoal is placed in a round-bottomed flask of hard glass filled with oxygen and closed with a tightly fitting rubber stopper. On heating the part of the flask in contact with the charcoal in a small bunsen flame, combustion is started and the flask then gently shaken (to prevent local heating of the glass) until combustion has ceased. When the flask has regained the temperature of the room it is opened under water. No gas escapes and no water enters, thus shewing the volume of gas to be unchanged. If the piece of charcoal is small and the oxygen in considerable excess, practically nothing but carbon dioxide is produced.

of extinguishing fires. Certain machines for this purpose either deliver carbon dioxide itself, or substances which liberate the gas, such as a solution of sodium bicarbonate. A brightly burning strip of magnesium continues to burn when plunged into the gas, with formation of magnesium oxide and black specks of carbon, an experiment sometimes quoted as proving that this gas contains both carbon and oxygen, and also as shewing that in some cases carbon dioxide can act as an oxidizing agent. Its density, which is 1·5 times that of air, permits of such spectacular experiments as pouring the gas from vessel to vessel almost as one would pour water, its presence or absence being demonstrated by the use of a burning candle. If collected at the bottom of a large open vessel, soap bubbles may be made to float on its invisible 'surface.' Many fatal accidents have been caused by this property through the collection of the gas at the bottom of wells and mines, in the large vats used in brewing, and in valleys in which lime burning is carried on.

The toxic action of carbon dioxide has been much exaggerated. The fatal effects caused by it are due to asphyxiation, that is, absence of oxygen, rather than to specific poisoning. Indeed it is probable that such direct poisoning as may be caused is due to traces of the monoxide present and not to the dioxide at all. Carbon dioxide liquefies at 0° C. under a pressure of 35 atmospheres. The earliest experiments on critical phenomena were made on this gas in 1869 by Andrews, who found the critical temperature to be 31° C., and the critical pressure 73 atmospheres.

When the compressed gas is allowed to expand, the fall in temperature may be sufficient to liquefy and even to solidify a portion of it. This solid, known as carbon dioxide snow, is obtained by tying a loose cloth or flannel bag over the nozzle of a cylinder of the compressed gas, and turning on the tap until the gas issues freely—a very wasteful process, but a ready one for obtaining low temperatures. This 'snow,' mixed with ether to assist evaporation, produces a temperature low enough to freeze mercury readily.

Carbon dioxide is much used in the making of artificial ice, the cooling being brought about by allowing the compressed gas to expand. It is afterwards compressed again and water cooled, little gas being lost in the process.

The coefficient of solubility in water is given by Bunsen as 1·798 at 0° C., and 1·002 at 15° C.; that is, water dissolves about its own volume of the gas at laboratory temperatures and under a pressure of one atmosphere *of carbon dioxide*. This can be driven out again completely by boiling (compare hydrochloric acid, p. 88).

Water saturated with the gas cannot be regarded as a simple aqueous solution of carbon dioxide. *Some* kind of chemical action is believed to take place, for the following reasons:

1. The solution is a much better conductor of electricity than pure water.

2. It possesses feebly acidic properties, turning litmus to a wine-red colour.

3. If a solution made under pressure be kept in a vessel fitted with a manometer, and from which neither gas nor liquid can escape, the pressure is observed to fall gradually, which is explained on the assumption that combination takes place between the gas and the water.

It is mainly on these grounds that a solution of carbon dioxide in water is said to contain a small quantity of carbonic acid, to which the formula H_2CO_3 is usually assigned. No such compound has ever been isolated, but as has already been seen, there are two well defined series of salts, carbonates and bicarbonates, which correspond to such a dibasic acid.

Solutions of alkaline hydroxides absorb carbon dioxide with formation of the carbonate:

$$2KOH + CO_2 = K_2CO_3 + H_2O.$$

This reaction is used for removing carbon dioxide from other gases which are not so absorbed. Air for instance, may be freed from all traces of carbon dioxide by passing it through a series of wash bottles containing a strong solution of potassium or sodium hydroxide. The same reaction is also employed for estimating the gas, both gravimetrically (Fig. 20) and volumetrically.

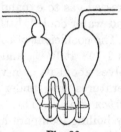

Fig. 20

The usual test for the presence of this gas is the formation of a white precipitate of insoluble carbonate when carbon dioxide comes in contact with clear lime water, or still better, baryta water:

$$Ba(OH)_2 + CO_2 = BaCO_3 + H_2O,$$

and since all carbonates or bicarbonates evolve the gas when acted on by dilute acids, this is one of the tests for carbonates in general. It should be noted that since even saturated lime water is a very dilute solution, it is not a good absorbent when more than traces of carbon dioxide are to be removed.

So-called 'soda' water and aerated waters in general, are made by dissolving carbon dioxide under pressure in water, and they usually contain but little else. In the manufacture, carbon dioxide snow provides a ready means of obtaining the required pressure, and the principle in the use of 'sparklets' is much the same. In bread-making the dough is caused to 'rise,' that is to become aerated and light, by the addition of yeast, which produces carbon dioxide from the flour by fermentation. Yeast is some-times replaced by 'baking powder,' usually sodium bi-carbonate, or sodium bicarbonate mixed with a solid organic acid such as tartaric acid, or more frequently potassium bitartrate, commonly called cream of tartar. The carbon dioxide is liberated either on warming, as in the case of sodium bicarbonate, or on moistening with water if an acid be used. Medicinal effervescing powders are of much the same nature, being usually mixtures of sodium bicarbonate and solid citric or tartaric acid. These interact when they come in contact with water, liberating carbon dioxide.

Perhaps the most important of the carbonates is that of sodium, since it is easily made, and almost all other metallic carbonates can be prepared from it. These are characterized as a class by very sparing solubility in water, and can be obtained by the addition of an aqueous solution of sodium carbonate to a solution of a salt of the required metal, thus:

$$Na_2CO_3 + MgSO_4 = MgCO_3 + Na_2SO_4,$$

the insoluble magnesium carbonate being filtered off, washed free from the soluble sodium sulphate produced, and thereupon dried.

Sodium carbonate is found naturally as a lacustrine deposit in many parts of the world. The natron lakes of Egypt supplied Europe for a long time with much of the sodium carbonate required for the two main industries depending upon it, namely, soap and glass making. The demand for sodium carbonate would soon have outstripped this source of supply under ordinary conditions, but the necessity for a more abundant supply was hastened and accentuated by political circumstances. One result of the battle of the Nile (Aboukir) in 1789 was to cut off the supply of sodium carbonate from France, at that time the most advanced country in the world in the matter of chemical industries. The difficulty was met by an offer from the French Government of a considerable reward for the discovery of a means of manufacture from materials to be had in France itself, and the outcome of this was the process known ever since by the name of the discoverer, Leblanc.

The Leblanc process is now practically obsolete, but in the course of a century or so it has enormously benefited not only France and England but the whole world. It has stimulated chemical industry in general to a degree difficult to exaggerate, and may be regarded as the parent of what are known as the 'heavy' chemical manufactures.

The process consisted in first converting common salt into sodium sulphate by the action of concentrated sulphuric acid (see Chap. III, p. 86). Sodium carbonate cannot be prepared directly by heating the sulphate with calcium carbonate, but if solid carbon be present the sulphate is reduced to the sulphide, from which sodium carbonate is easily obtained by interaction with calcium carbonate. The sodium sulphate (salt cake) was mixed with coke and limestone and heated strongly (black ash process). The reactions which occur may be represented by the equations:

$$Na_2SO_4 + 4C = Na_2S + 4CO,$$
$$Na_2SO_4 + 2C = Na_2S + 2CO_2,$$

and $\qquad Na_2S + CaCO_3 = Na_2CO_3 + CaS,$

the double decomposition, begun in the furnace, was completed in the third stage, the lixiviation with water, from which the final product crystallized as the decahydrate $Na_2CO_3 . 10H_2O$, or ordinary washing soda of commerce.

A modern process, which can be considered as great an advance on the Leblanc process as the latter was on recovering sodium carbonate from the natron deposits of

Egypt, is the Solvay or ammonia-soda process. This depends for its success on the smaller solubility of the bicarbonate of sodium than that of the normal carbonate, which is an inversion of the usual relationship between the two classes of salts, since in almost all other cases the bicarbonate has the greater solubility.

The process can be illustrated in the following way. A litre flask is half filled with an aqueous solution saturated with both common salt and ammonia. The flask is closed with a stopper, through which passes a tube connected with a Kipp's apparatus delivering carbon dioxide under slight pressure. The absorption of the gas is slow but continues for a long time. After a few hours a precipitate of sodium bicarbonate separates, and continues to increase for several days. The reaction which takes place is thus represented:

$$NaCl + NH_3 + H_2O + CO_2 = NaHCO_3 + NH_4Cl.$$

There are few waste products in this process. The ammonia is recovered with but slight loss from the ammonium chloride produced, by heating the latter with lime (p. 148). The lime is obtained by heating limestone as already described, the carbon dioxide produced at the same time being used as above. Theoretically therefore the only necessary raw materials, which are very cheap, are salt, limestone, and fuel, while the only bye-product unused is calcium chloride—a waste material in almost all chemical industrial processes.

The method is sometimes further cheapened by using a 'substituted' ammonia, trimethylamine, $N(CH_3)_3$, instead of ammonia. This is itself a waste product of the beet sugar industry, and has the additional advantage of being more effective than ammonia, since its solubility in cold water is infinite (see p. 35).

It will be noted that the first Solvay product is the

bicarbonate. The normal carbonate is easily obtained from this by gentle heating, and the carbon dioxide thus liberated is used again in the first operation described above. Potassium bicarbonate cannot be manufactured by the Solvay process, owing to the small difference in solubility between it and the normal carbonate.

Carbon Monoxide. The technical manufacture of this gas on a large scale has been previously described (p. 116). For experimental purposes fairly pure carbon monoxide may be made by passing the dioxide over red hot charcoal, and removing traces of unchanged dioxide by absorption in potash. It is more usually obtained by withdrawal of the elements of water from certain organic acids by means of concentrated sulphuric acid. A quantity of oxalic acid is placed in a flask and covered with sulphuric acid. The gas which is evolved on gently warming is passed through one or more wash-bottles containing potassium or sodium hydroxide, and collected over water in the usual way.

$$H_2C_2O_4 = H_2O + CO_2 + CO.$$

Sodium formate may be used in place of oxalic acid, in this case no carbon dioxide being produced. The formic acid at first liberated is immediately decomposed thus:

$$CH_2O_2 = H_2O + CO.$$

Carbon monoxide easily reduces many metallic oxides at a red heat. With half its volume of oxygen it forms a highly explosive mixture, the volume of the product being equal to the original volume of the carbon monoxide, and from this the molecular formula CO can be deduced. It is not absorbed by aqueous potash, but *fused* potash slowly combines directly with the gas, producing potassium formate:

$$CO + KOH = HCOOK.$$

Although the gas is practically insoluble in water and does

not combine directly with it, the above equation indicates a method of bringing about the combination indirectly. Carbon monoxide in presence of water has no acidic properties, and yet this reaction indicates that it may be regarded as the anhydride of formic acid, HCOOH.

A solution of cuprous chloride, either in concentrated hydrochloric acid or in aqueous ammonia, slowly absorbs the gas, and this is the usual method of estimating carbon monoxide in gas analysis. It combines directly with the red colouring matter of blood, forming carboxy-haemoglobin. This compound is so easily identified (and estimated) spectroscopically that it offers one of the readiest and most sensitive methods of detecting even small traces of the gas. Certain metals, when in a finely divided condition, combine with carbon monoxide, forming compounds known as metallic carbonyls, the best known of which is nickel carbonyl, a volatile liquid having the formula $Ni(CO)_4$. These compounds all dissociate very readily on heating. Nickel is obtained from its ores in very pure condition by these reactions. Cobalt usually accompanies nickel in its ores, and since it does not form a carbonyl under the conditions employed, the nickel obtained by this method is free from cobalt.

When a mixture of equal volumes of carbon monoxide and chlorine is exposed to sunlight, direct union takes place slowly with formation of carbonyl chloride, or phosgene, $COCl_2$. The combination takes place more rapidly in presence of a catalyst, and for this purpose certain forms of charcoal are used. The gas is easily liquefied, and is usually preserved in that form in sealed glass vessels under slight pressure. It is a very poisonous gas, rapidly leading to cardiac failure when inhaled. It readily reacts with alkalis, forming chlorides and carbonates:

$$COCl_2 + 4KOH = 2KCl + K_2CO_3 + 2H_2O,$$

9-2

and also, though more slowly, with water and with alcohol. It is sometimes found in chloroform which has been much exposed to light, as the product of atmospheric oxidation:

$$2CHCl_3 + O_2 = 2COCl_2 + 2HCl.$$

Hydrides of Carbon. When the mud at the bottom of a river or pond is disturbed, bubbles of gas usually make their appearance. If these be examined it will be found that the gas is inflammable, and that the products of its combustion are carbon dioxide and water. It consists for the greater part of methane, or marsh gas, CH_4, and occurs naturally where vegetation decomposes in presence of moisture. The gas from this source is sometimes spontaneously inflammable, the flickering flames which appear being known as Will o' the wisp. The spontaneous inflammability is due to traces of other compounds, probably phosphine for the greater part. Natural gas, which issues from the ground under very considerable pressure in some parts of the world, usually contains a high percentage of methane. It is also found in coal mines, especially where soft coal is obtained, and is known to the miners as 'fire damp.' Owing to the highly explosive nature of a mixture of the gas with air, it has been a cause of very serious accidents. The distillation of coal produces much methane, ordinary coal gas frequently containing as much as 50 per cent. of it, the rest being mainly hydrogen.

When a mixture of three volumes of hydrogen with one volume of carbon monoxide is passed through a tube containing finely divided nickel heated to about 300° C., methane and steam are produced:

$$3H_2 + CO = CH_4 + H_2O.$$

It may also be prepared by the reduction of methyl iodide

CH_3I (p. 104) this being usually carried out by means of a zinc-copper couple:

$$CH_3I + H_2 = CH_4 + HI.$$

A somewhat impure product may be prepared from sodium acetate and soda lime. An intimate mixture of the dry materials is usually heated in a tube of iron, copper, or hard glass, and the gas collected over water:

$$CH_3COONa + NaOH = Na_2CO_3 + CH_4.$$

Soda lime is a mechanical mixture obtained by slaking quicklime with a solution of caustic soda and heating to granulation. It is used instead of sodium hydroxide in cases where fusion is undesirable.

Methane is colourless and practically insoluble in water. It burns with but little luminosity, and considerable evolution of heat:

$$CH_4 + 2O_2 = CO_2 + 2H_2O.$$

As will be seen from the equation, one volume of marsh gas requires two volumes of oxygen (or ten volumes of air) for complete combustion. A mixture of air and methane which contains less than five per cent. of the latter gas is not explosive at ordinary pressure.

Methane is incapable of combining *directly* with any other substance. Whenever it undergoes a chemical change, one or more of the hydrogen atoms is removed from the molecule and replaced by something else. When a mixture of equal volumes of methane and chlorine is exposed to light, a slow reaction takes place; hydrogen chloride is produced, together with a compound derived by the replacement of one of the hydrogen atoms of the methane molecule by an atom of chlorine:

$$CH_4 + Cl_2 = CH_3Cl + HCl.$$

The first product of this reaction is a modified methane; chlor-methane, or methyl chloride. The process, known as *substitution*, continues if sufficient chlorine be present, until all the hydrogen of the methane molecule has been replaced by chlorine, thus:

$$CH_3Cl + Cl_2 = CH_2Cl_2 + HCl.$$
$$CH_2Cl_2 + Cl_2 = CHCl_3 + HCl.$$
$$CHCl_3 + Cl_2 = CCl_4 + HCl,$$

the products being di-chlor-methane, tri-chlor-methane or chloroform, and finally tetra-chlor-methane or carbon tetrachloride.

It should not be assumed that each stage in these substitutions is completed before the next begins. Such is far from being the case. As each atom of hydrogen is replaced by chlorine, the chlorinated molecule becomes progressively less and less resistant to the change, the last atom of hydrogen in the molecule being the most easily replaced. If therefore equal volumes of methane and chlorine be mixed and exposed to light, when all action has ceased the vessel will contain *all* the above chlor-methanes, in varying quantities, together with some unchanged methane.

These reactions are typical of the changes which take place between *saturated* compounds, like methane and the other members of the paraffin series, and chlorine or bromine. The reaction is sometimes used for the preparation of hydrogen bromide, but as will be seen from the equations the process is a wasteful one, since for every atom of bromine converted into hydrogen bromide an atom of bromine is lost in the useless brom-paraffin. The operation is carried out by dropping bromine slowly on to gently warmed paraffin oil or vaseline. The brom-paraffin remains behind in the flask, while the hydrogen bromide, freed from uncombined bromine by passage over phosphorus, is either dissolved in water or collected over mercury. A suitable apparatus is illustrated in Fig. 16.

A hydrocarbon of a different type may be obtained by removing the elements of water from ordinary alcohol:

$$C_2H_5OH = C_2H_4 + H_2O.$$

This may be done by strongly heating a mixture of alcohol and concentrated sulphuric acid, and collecting the gas evolved over water in the usual way. A more convenient method, when large quantities of the gas are required, is found in passing the vapour of alcohol over hot charcoal which has previously been impregnated with phosphoric acid, when the alcohol is decomposed into the hydrocarbon and water vapour. The *ethylene* thus obtained is the first member of a series of compounds of carbon and hydrogen of the general formula C_nH_{2n}, known as olefines from their property of forming oily products with chlorine.

Ethylene burns with a much more luminous flame than methane, and requires three times its volume of oxygen for complete combustion:

$$C_2H_4 + 3O_2 = 2CO_2 + 2H_2O.$$

The members of this series differ from the paraffins in that they combine *directly* with chlorine, bromine, and hydrogen, forming additive products, and are therefore said to be *unsaturated*. If ethylene be passed into bromine direct combination takes place, with the formation of a heavy colourless liquid, ethylene dibromide:

$$C_2H_4 + Br_2 = C_2H_4Br_2.$$

Similarly, on mixing equal volumes of ethylene and chlorine the corresponding ethylene dichloride, $C_2H_4Cl_2$, is produced. The combination with hydrogen does not take place so readily, but may be brought about by passing a mixture of ethylene and hydrogen over a catalyst, such as finely divided nickel, at a moderate temperature, when the second member of the paraffin series, *ethane*, C_2H_6, is formed.

By boiling ethylene dibromide with an alcoholic solution of caustic potash, the elements of hydrogen bromide are removed from the compound, and *acetylene*, a hydrocarbon of a third type, is produced:

$$C_2H_4Br_2 + 2KOH = C_2H_2 + 2KBr + 2H_2O.$$

It is more readily obtained, though in a less pure condition, by the action of water on calcium carbide, CaC_2, a compound made by heating a mixture of lime and coke to a high temperature in an electric furnace. The carbide reacts readily with water, yielding acetylene and slaked lime:

$$CaC_2 + 2H_2O = C_2H_2 + Ca(OH)_2,$$

and this process provides a very cheap illuminating gas. Two volumes of acetylene require five volumes of oxygen for complete combustion,

$$2C_2H_2 + 5O_2 = 4CO_2 + 2H_2O,$$

and since acetylene is an endothermic compound, it burns with a very hot flame*. The oxy-acetylene flame finds considerable application in steel welding.

Compounds of the acetylene series, C_nH_{2n-2}, shew unsaturation in a more marked degree than ethylene and its homologues. They can combine directly with chlorine and bromine to form compounds of the type $C_2H_2Cl_4$ and $C_2H_2Br_4$, and, under the influence of catalysts, with hydrogen, to produce firstly ethylene and then ethane.

---

* This is shewn by the following thermo-chemical equation:

Heats of formation: $C_2H_2 = -58K$; $CO_2 = 97\cdot6K$; H_2O (steam) $= 58K$.

$$2C_2H_2 + 5O_2 = 4CO_2 + 2H_2O + E$$
$$2 \times -58 \qquad 4 \times 97\cdot6 \quad 2 \times 58,$$
$$E = (2 \times 58) + (4 \times 97\cdot6) - (2 \times -58)$$
$$= \quad 116 \quad + \quad 390\cdot4 \quad + \quad 116$$
$$= \quad 622\cdot4K.$$

Therefore the heat of combustion of one gramme-molecule of acetylene is $311\cdot2K$.

When unsaturated hydrocarbons of the ethylene or acetylene series are agitated with a dilute solution of potassium permanganate made alkaline with sodium carbonate, the solution is decolorized, and a precipitate of hydrated manganese dioxide appears. This is known as von Baeyer's test for unsaturation. They decolorize bromine water, and are dissolved by concentrated sulphuric acid.

When acetylene is passed into an ammoniacal solution of copper hydroxide a reddish precipitate is produced. A white precipitate is obtained when the gas is similarly absorbed by a solution of silver oxide in aqueous ammonia. Both these compounds are very explosive when dry, and this renders examination of them difficult, and therefore little is known about their molecular composition. They are most probably carbides of the metals, and may be represented by the empirical formulae CuC and AgC. Acetylene is once more evolved when these compounds are warmed with hydrochloric acid, a reaction which often occurs between metallic carbides and acids.

The production of copper carbide may be made a very sensitive test for acetylene in the following way. To a few cubic centimetres of a dilute solution of copper sulphate, ammonia is added until the precipitate at first formed is redissolved. To this solution hydroxylamine hydrochloride is added until the liquid is reduced to the colourless cuprous condition. If the resulting solution be then poured into a jar containing coal gas and well shaken, the liquid becomes reddish pink in colour; on standing, the finely divided copper carbide slowly settles. A mere trace of acetylene can be detected in this way.

CHAPTER V

NITROGEN AND ITS COMPOUNDS

THE most convenient source of nitrogen, when required in large quantities, is the atmosphere, which contains about 79 per cent. by volume. When liquid air is allowed to evaporate the first portions to boil off are almost pure nitrogen, and the last portions almost equally pure oxygen. Air is liquefied on a large scale for the purpose of separating these two gases for technical uses, and this is practically the only method now employed for the commercial supply of oxygen and nitrogen.

When air is passed over red hot copper the whole of the oxygen is removed if contact with the copper be sufficiently prolonged. This method also has been applied for the technical preparation of nitrogen.

When ammonia gas is passed over red hot copper oxide the latter is reduced to metallic copper, with formation of water vapour and free nitrogen. If a mixture of air and ammonia be used, nitrogen is obtained from both sources, the copper being apparently unchanged if the ammonia be in slight excess. The residual ammonia is removed by collecting the nitrogen over slightly acidulated water. It is easy to see how the copper functions here, and this reaction is sometimes quoted as an instance of the cyclic nature of catalytic action (p. 70).

If a piece of phosphorus be ignited in a closed vessel of air, combustion continues until *the whole* of the oxygen is removed by union with the phosphorus, if the latter be in excess (Fig. 21). Few substances are capable of doing this. A burning candle plunged into a jar of air is extinguished long before all the oxygen is used up, and the same thing takes place with burning sulphur, or charcoal, or alcohol. Phosphorus can indeed remove all the oxygen from air without actually taking fire, and is so used for estimating the relative volumes of oxygen and nitrogen in the atmosphere.

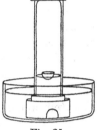

A graduated cylinder nearly filled with air is inverted over water, the water levels adjusted and the volume of air then read, the temperature and pressure being noted. A stick of phosphorus attached to a length of wire is then

Fig. 21

introduced into the cylinder above the water level, and the apparatus kept in a cool place so that the phosphorus may not take fire. When there is no further diminution in the volume of the gas the phosphorus is withdrawn, the water levels again adjusted, and the volume, temperature, and pressure again read. The disappearing volume is that of the oxygen; the residual gas, nitrogen. Since in the above experiment other constituents of the atmosphere are ignored, the result is only approximately accurate.

The laboratory preparation of nitrogen is made by boiling an aqueous solution of ammonium nitrite, which decomposes on heating:

$$NH_4NO_2 = N_2 + 2H_2O.$$

The ammonium nitrite may be replaced by a mixture of the more accessible materials ammonium chloride and sodium nitrite, which on being boiled together react thus:

$$NH_4Cl + NaNO_2 = NaCl + 2H_2O + N_2,$$

the gas being collected over water.

When organic compounds containing nitrogen are heated in presence of copper oxide the nitrogen is liberated, while the carbon and hydrogen are oxidized to carbon dioxide and water. The nitrogen is made to pass over a length of red hot copper gauze to remove oxygen from any oxide of nitrogen which may be present, and the gas is then collected over a solution of caustic potash, which absorbs the carbon dioxide. As the whole of the nitrogen contained in the substance can be set free and collected in this way, the method is used for estimating the percentage of nitrogen in organic compounds generally. When chlorine is passed through an aqueous solution of ammonia, nitrogen is liberated, and ammonium chloride left in solution:

$$8NH_3 + 3Cl_2 = N_2 + 6NH_4Cl.$$

Although this is important as a reaction, it is not a convenient way of preparing the gas.

Nitrogen is characterized by its negative properties. It is neither combustible nor a supporter of combustion in the ordinary meaning of the terms. It is only very slightly soluble in water. It is not poisonous, but it can do nothing to support life. It functions in the atmosphere as a diluent, and when breathed appears to take no part in the animal economy, leaving the lungs as it enters, unchanged in nature and amount. There are, however, micro-organisms which are able to bring free nitrogen into combination. Certain bacteria found in the root nodules of peas, beans, and Leguminosae generally, increase the amount of nitrates in the soil in which the plants grow, and since their only source of nitrogen is the atmosphere, they must be regarded as agents in the fixation of uncombined nitrogen.

The bringing of uncombined nitrogen into combination is a matter of the greatest importance, since, as will be seen later, the food supply of the world will ultimately depend upon it. As might be supposed from its inert nature, the gas does not readily combine with any element or compound. As long ago as 1784, Cavendish shewed (Fig. 22) that when a mixture of nitrogen and oxygen is subjected to the prolonged action of electric sparks, union takes place slowly, the resulting products dissolving in water to yield an acid solution. Priestley had previously (1779) observed that air acquires an acid reaction upon being sparked, but it is doubtful whether he understood the significance of his experiments.

This discovery was not seriously investigated until more than a century later, when Lord Rayleigh returned to it, for a reason quite unconnected with the production of nitric acid, and which will appear later. Instead of the

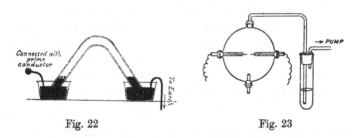

Fig. 22   Fig. 23

frictional machine used by the earlier experimenters, he employed a powerful electric arc within a glass globe containing a mixture of oxygen and nitrogen. He absorbed the products as they were formed in a solution of caustic soda, and in this way obtained a yield corresponding to 46 grm. of nitric acid per kilowatt-hour. A simplified form of the apparatus is shewn in Fig. 23. Some years previously Crookes had drawn attention to the rapidly diminishing world supply of natural nitrates, and had indicated the seriousness of this in connection with the wheat crops. McDougall and Howliss made the earliest attempt to put the discoveries of Priestley, Cavendish, and Rayleigh on a commercial basis, their operations being carried on in

Manchester in 1899, but neither their efforts, nor those of Bradley and Lovejoy at Niagara Falls, were successful in producing nitrates which could compete in price with the supply still accessible from the Chile deposits (p. 160), and it was not until 1903 that a successful commercial process was invented. This process, known by the names of its discoverers, Birkeland and Eyde, has developed considerably in Norway, where hydro-electric power is cheap, and the enormously increased demand for nitric acid during the years 1914–1918 further so stimulated research in this direction, that in the result much of the nitric acid now used is synthetic in origin.

In the Birkeland-Eyde process atmospheric oxygen and nitrogen unite under the influence of the electric arc. The latter is produced between hollow water-cooled copper electrodes, and made to assume a fan or disc shape, some six feet in diameter, by the use of powerful electro-magnets placed at right angles to the arc. Since the reaction:

$$N_2 + O_2 = 2NO$$

is endothermic (and reversible), in accordance with Le Chatelier's principle a high temperature must be employed, in this case approximating to 3000° C. Equilibrium between the gases is reached in a few seconds, when about 1 to 2 per cent. of nitric oxide is present. In order to diminish the rate at which the reverse action,

$$2NO = N_2 + O_2,$$

takes place, the gases must be cooled quickly, and at about 700° C. this reversion becomes practically negligible. The temperature of the gases is further lowered to about 50° C. by passage through large water-cooled aluminium pipes, where the combination of the nitric oxide with oxygen takes place:

$$2NO + O_2 = 2NO_2,$$

this reaction being completed in about two minutes. Incidentally, in the cooling of the gases more steam is generated than suffices for the needs of the whole works.

The conversion of nitric oxide into nitric acid and nitrates is attained by passing the gases, now consisting of nitrogen, oxygen, and nitrogen peroxide, through a series of scrubbers. These are enormous granite towers packed with quartz fragments to give a large surface, over which a thin stream of water percolates. The reactions which take place in presence of excess of oxygen may be represented thus:

$$4NO_2 + 2H_2O + O_2 = 4HNO_3.$$

About 85 per cent. of the nitric peroxide present is thus directly converted into nitric acid. The greater part of the remainder is recovered by passing the gases issuing from the first series of scrubbers through similar but smaller ones, down which a solution of sodium carbonate trickles. Sodium nitrite and sodium nitrate are produced by the reaction:

$$2NO_2 + Na_2CO_3 = NaNO_3 + NaNO_2 + CO_2.$$

Owing to the expense and danger entailed in the carriage of nitric acid, it is often marketed in the form of ammonium or calcium nitrate (Norge saltpetre). The former is obtained by neutralizing the acid with ammonia, made by a process described later; the latter by neutralizing it with limestone. In both cases the neutral solutions are evaporated to dryness.

At high temperatures nitrogen reacts with calcium carbide to produce calcium cyanamide and free carbon. This provides a commercial method for the synthetic production of ammonia, and therefore of nitrogen compounds generally. Calcium carbide is heated in an atmosphere of nitrogen (from liquid air) to a temperature of about 1100°C.

The reaction, which may be represented by the equation

$$CaC_2 + N_2 = CaCN_2 + C$$

is exothermic, and when once started proceeds without further external heating. The mixture of calcium cyanamide and carbon contains also a little unchanged carbide. The latter compound is decomposed by adding small quantities of water, which, as has already been shewn, produces calcium hydroxide. After this treatment the mass is ground to a fine powder and sent into the market under the name of nitrolim. In this form it is used largely in agriculture as a fertiliser, but considerable quantities of it are also used for the production of ammonia. Cold water reacts slowly with calcium cyanamide, but at higher temperatures decomposition takes place readily, with formation of ammonia and calcium carbonate:

$$CaCN_2 + 3H_2O = CaCO_3 + 2NH_3.$$

The operation is carried out by treating calcium cyanamide in autoclaves in presence of a little soda, with steam under pressure. The action may be illustrated by heating a little of the compound almost to redness in a test tube and allowing a few drops of water to fall upon it, when the evolution of ammonia is obvious.

A further synthesis of ammonia, which has developed into considerable importance in recent years, is known by the name of the inventor, Haber. In the Haber process the direct union of nitrogen and hydrogen is brought about by the influence of suitable catalysts.

$$N_2 + 3H_2 = 2NH_3.$$

Since ammonia is an exothermic compound, theoretically the temperature should be kept low. Experience shews however that at temperatures below 500° C. the velocity of the reaction is too slow to be technically useful. According to the above equation, the volume of the ammonia is only one half of the sum of the volumes of the hydrogen and nitrogen from which it is produced; increase of pressure should therefore, according to Le Chatelier's theorem, increase the yield of ammonia, and actually a pressure of 200 atmospheres is employed. The optimum temperatures and pressures are not yet established, and variations in both are to be found in the works engaged in these operations. The catalyst used also differs in various localities. It is generally a finely divided metal; iron, osmium, molybdenum, and uranium having been suggested. In each works however the nature of the catalyst is a jealously guarded secret.

Nitric acid is manufactured from ammonia on the technical scale by an oxidation process devised by Ostwald and Brauer. The ammonia, mixed with air, is passed over a heated catalyst, platinum being the most efficient. Water and nitric oxide are first formed, the latter being converted into nitric acid in a manner similar to that employed in the Birkeland-Eyde process. The most efficient works convert more than 90 per cent. of the ammonia used into nitric acid.

Nitrogen combines directly with certain metals, e.g. magnesium, to form metallic nitrides. This reaction however has no technical value for the fixation of nitrogen. It can be shewn experimentally by holding the burning end of a piece of magnesium ribbon in a crucible in such a way as to check the supply of oxygen without actually extinguishing the flame. The white product is mainly magnesium oxide, but the presence of some nitride, Mg_3N_2, can be shewn by warming the material with dilute sodium hydroxide, when the evolution of ammonia becomes unmistakable.

An active modification of nitrogen can be produced by jar discharges of electricity through ordinary nitrogen (Strutt, 1911). This is analogous to the production of ozone from oxygen, the active form of nitrogen being considered to be an allotropic modification of the element. While the atomicity of the two forms of oxygen is well established, that of active nitrogen cannot yet be regarded as beyond doubt, though it has been suggested, in view of its extremely active nature, that it may be monatomic. It seems to attack many substances, including most of the elements, but the quantities hitherto obtained have been so small that the nature of its action is largely conjectural.

In the course of a series of experiments carried out in 1894, Lord Rayleigh found that the density of nitrogen obtained from the air was always greater than that of the gas prepared from its compounds, and that this difference maintained a constant value. The respective numbers obtained were 1·25718 grm. per litre for atmospheric nitrogen, and 1·25107 for the gas obtained from its compounds. This result seemed to point to the presence of a hitherto unknown gas in the atmosphere, somewhat heavier than nitrogen.

In his experiments on the sparking of a mixture of air and oxygen referred to above, Cavendish had noticed that there was always a residue of gas which could not be made to combine with oxygen under the influence of prolonged sparking, and that this residue amounted to about 1/120 part of the nitrogen present. Recalling this observation, Rayleigh repeated Cavendish's experiment on a larger scale and with more effective apparatus, and came to the conclusion that the residue noticed by Cavendish really did consist of a new gas, which further investigation has shewn to be a mixture of several gases, of a type the existence of which was hitherto unsuspected. With the assistance of

Ramsay, other methods of removing the nitrogen were tried. Air was passed over red hot copper to absorb oxygen, and the remaining gas passed over red hot magnesium which, as has been seen, combines with nitrogen, the new gas remaining unaffected. This residual gas has since been separated into five distinct individuals, argon, helium, neon, krypton, and xenon. They are characterized by the monatomic structure of their molecules, and their inability to enter into any kind of combination whatever. They form what is known as the inert or non-valent group of elements. It is interesting to note that in the Haber process these gases accumulate to such an extent in the spent products as to necessitate periodical removal by blowing off. The non-valent gases constitute about one per cent. of the volume of the atmosphere. They are now obtained in the first place by the fractional distillation of liquid air. Argon is used for filling electric filament lamps, one third of an atmosphere of this gas considerably reducing the rate of blackening of the lamps which occurs when vacuous globes are used. Glow (non-filament) lamps, containing neon under small pressure, emit a beautiful red light.

Ammonia is one of the products of decaying or fermenting nitrogenous organic matter. It occurs in herring brine, together with substituted ammonias, chiefly methylamine, NH_2CH_3, and its odour is noticeable in places where the products of animal metabolism are allowed to accumulate. Large quantities of it are obtained in the form of ammonium sulphate from the liquors of gas works and from the spent gases of furnaces. These gases are 'scrubbed' with water, in which the ammonia dissolves, and the liquors so obtained are distilled into dilute acid, either hydrochloric or sulphuric, and the neutralized product evaporated down. The South Metropolitan Gas Company recovers six hundredweight of ammonia, corresponding to 1·2 tons of ammonium sulphate, from every hundred tons of coal distilled. Certain kinds of nitrogenous organic matter, such as albumin, hair, feathers, and leather, evolve ammonia when heated with soda lime, and the gas can also be detected when such substances are burnt.

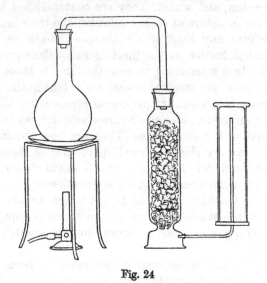

Fig. 24

The laboratory preparation is usually carried out by mixing any ammonium salt, generally the chloride, with an excess of quicklime and gently warming.

$$2NH_4Cl + CaO = CaCl_2 + H_2O + 2NH_3.$$

Since water is one of the products of the reaction, the excess of quicklime functions as a very convenient drying agent. Further drying is effected by passing the gas down a glass vessel filled with lumps of quicklime and known as a 'lime tower' (Fig. 24). The usual drying agents, concentrated sulphuric acid and calcium chloride, cannot be used since they both combine with ammonia. When the gas itself is required it may be collected over mercury.

The exhaustive drying of ammonia for special purposes is carried out by leaving the gas in contact with phosphorus pentoxide, P_2O_5, for a long time. Since the latter compound unites with water to form phosphoric acid, and this in turn combines with ammonia to form ammonium phosphate, it is plain that such drying is attended with a loss of gas. To minimise this loss, a preliminary drying with quicklime should always be made.

Ammonia is a colourless gas of well-known odour and is much lighter than air. It neither burns nor supports combustion under ordinary conditions, but can be made to burn in an atmosphere of oxygen, when it gives a livid yellowish flame. As the critical temperature of ammonia is 130° C. it is easily liquefied by pressure at ordinary temperatures, a pressure of rather more than six atmospheres being required at 10° C. This renders it particularly suitable for use in freezing machines for the production of ice on a large scale, and for maintaining low temperatures in cold storage rooms. Liquid ammonia has a specific gravity of 0·63 at 0° C. and boils at − 33·5° C. under ordinary atmospheric pressure.

The gas is extremely soluble in water. The coefficient of solubility at 0° C. is difficult to determine with accuracy, and widely differing values have been given by various investigators. Raoult found that one volume of water at 0° C. dissolved 1305 volumes of ammonia under atmospheric pressure, though rather lower figures have been given by other experimenters. Water saturated with the gas at 14° C. under atmospheric pressure contains about 800 times its volume of ammonia, corresponding to 36 per cent. of the gas by weight. This solution has a specific gravity of 0·884 and is the *liquor ammoniae fortissimus* of commerce. As may be verified by a simple calculation, the volume of the water has been nearly doubled by the absorption of the gas.

The relation between the pressure and the volume when ammonia is dissolved in water is not strictly in agreement with Henry's law, the gas being more soluble at high pressures than the law demands. The solution has electrolytic properties, though it is by no means a good conductor of electricity. It has also an alkaline reaction, and to some extent resembles the soluble metallic hydroxides, such as those of sodium and potassium, in its power of precipitating certain insoluble metallic hydroxides from solutions of their salts. It is for these reasons that ammonia is considered to unite with water to some slight extent, forming ammonium hydroxide, NH_4OH. Since ammonia can be removed completely from its solution by boiling, this reaction is reversible,

$$NH_3 + H_2O \rightleftharpoons NH_4OH.$$

It should be clearly understood that ammonium hydroxide has never been isolated, and if it exists at all, it only does so *in solution*, and to a very slight extent. In this respect ammonium hydroxide resembles carbonic acid (q.v.).

Molecular Composition of Ammonia. Hofmann's experiment, which shews that ammonia consists of three volumes of hydrogen united with one volume of nitrogen, is carried out as follows. A tube about one metre long, closed at one end, is filled with chlorine under atmospheric pressure, and fitted at the open end with a tap funnel passing through a rubber stopper and containing a concentrated solution of ammonia (Fig. 25). The tube is held vertically, and on admitting a few drops of the liquid into it by a rapid opening and closing of the tap an energetic reaction takes place, with formation of ammonium chloride and a disappearance of some of the chlorine. When the tube has cooled somewhat, more ammonia solution is passed in, and the tube again allowed to cool. Finally the funnel is filled with water and the tap fully opened. Water flows into the tube until the pressure inside is in approximate equilibrium with that of the atmosphere, when it is observed that the liquid fills two-thirds of the tube, the remaining one-third being occupied by the nitrogen liberated in the reaction.

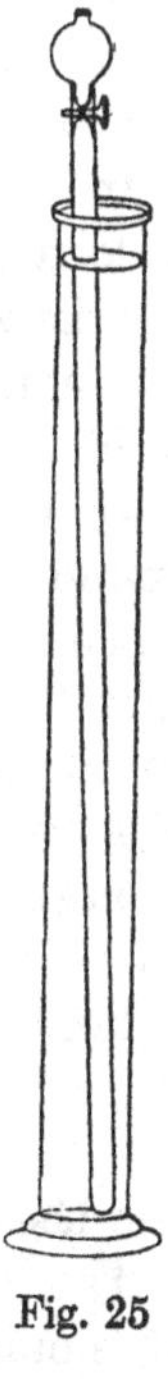

Since, as has been shewn, chlorine and hydrogen unite only in equal volumes, the volume of hydrogen present in the ammonia decomposed is equal to the volume of the tube, while the volume of the nitrogen liberated is one third of this. Hence, in ammonia, three volumes of hydrogen are combined with one volume of nitrogen. Since both nitrogen and hydrogen molecules are diatomic, the molecule of ammonia must be $(NH_3)_x$. It should be noted that this experiment gives no indication of the value of x, but since the specific gravity of ammonia is 8·5 in terms of hydrogen as unity, its molecular weight must be 17, and since the atomic weight of nitrogen is 14, x must equal unity.

Fig. 25

The value of x can be shewn in another way. As has been seen, nitrogen and hydrogen unite under the influence of electric sparks to form ammonia, though equilibrium is reached when only a very small proportion of the gases have so united. On the other hand, when ammonia is sparked it decomposes into nitrogen and hydrogen, though this decomposition is never complete, however long the sparking is continued. It follows therefore that if two

tubes be taken, the one containing ammonia and the other
a mixture of three volumes of hydrogen with one volume
of nitrogen, and sparks passed through each of them until
no further change takes place, the composition of the gas
in the two tubes will be identical, and will consist mainly
of mixed nitrogen and hydrogen, together with a small
percentage of ammonia.

In an actual experiment, 10 c.c. of dry ammonia were
collected over mercury in a graduated sparking tube.
Sparks were passed for a few minutes and the tube allowed
to regain room temperature. The volume of gas was found
to be 19 c.c. On admitting a few drops of dilute sulphuric
acid above the mercury, in order to dissolve the unchanged
ammonia, the volume of the residual gas became 18 c.c.
Since 1 c.c. of ammonia remained unchanged, it follows
that 9 c.c. of ammonia had yielded 18 c.c. of gas on decomposing. Applying Avogadro's theorem, two molecules
of ammonia yield four molecules of mixed gas, which has
been shewn above to consist of three molecules of hydrogen
and one molecule of nitrogen. The only equation consistent with these experimental results is therefore:

$$2NH_3 \rightleftharpoons N_2 + 3H_2,$$

and the molecular formula of ammonia must be NH_3.

The composition of the gas obtained by sparking
ammonia and dissolving away any of the latter remaining
undecomposed, can also be verified in another way. Excess
of oxygen is added to a measured volume of the mixed
gases, and a spark passed through the mixture. An explosion follows, all the hydrogen being converted into
water which of course condenses. The excess of oxygen is
then removed, either by absorption in alkaline pyrogallol,
or by introducing a piece of phosphorus attached to a wire
and allowing the phosphorus to remain in contact with the

gases until no further diminution of volume takes place. The phosphorus is then withdrawn and the volume of the residual gas, which consists of nitrogen only, is read. The volume of the hydrogen is obtained by difference, and may be checked by calculation, since the gas which disappears on explosion consists of two volumes of hydrogen and one volume of oxygen.

Valency. The relationship between the atomic weight of an element and its equivalent weight is shewn in the equation:

$$\text{Atomic weight} = \text{equivalent weight} \times n,$$

where n is a small integral number rarely exceeding 5, and never exceeding 8. This integer is sometimes defined as the *valency* of the element. Valency may also be considered from a different point of view, as representing the capacity of an element for saturation with hydrogen atoms. From this standpoint oxygen is considered to be bivalent, since the molecular formula of water vapour is H_2O, the atom of hydrogen being the standard of valency. Chlorine is considered to be univalent, as may be seen from the molecule of hydrogen chloride, HCl. It is frequently convenient to represent the valency of an element graphically by 'bonds.' Thus water and hydrogen chloride may be written H—O—H and H—Cl respectively. Experience shews that the valency of some elements is invariable, as is the case with sodium and hydrogen, whereas that of others, such as iron, does vary. The molecules of ferrous and ferric chlorides at high temperatures are represented by the formulae $FeCl_2$ and $FeCl_3$ respectively; in other words, the valency of iron in ferrous salts is two, while in ferric salts it is three. In such cases there is a corresponding difference in the equivalent weights of the element in the two kinds of compound, the above equation always being satisfied. The oxidation of ferrous chloride to ferric chloride by chlorine may be represented by a graphical equation as follows:

$$2Fe\begin{matrix} Cl \\ \\ Cl \end{matrix} + \begin{matrix} Cl \\ | \\ Cl \end{matrix} = 2Fe\begin{matrix} Cl \\ -Cl \\ Cl \end{matrix}$$

Although the recognition of the variable nature of valency has been necessary in order to explain many phenomena, this has led to difficulties in certain comparatively simple cases. There is for instance more than one way of expressing the molecular constitution of hydrogen peroxide. If both atoms of oxygen be regarded as bivalent, the compound may be written thus:

$$H-O-O-H.$$

If both oxygen atoms are assumed to be quadrivalent (and oxygen probably is quadrivalent in certain organic compounds) the formula would become

$$H-O\equiv O-H.$$

A structure, suggested by Traube, gives expression to the readiness with which hydrogen peroxide decomposes into water and oxygen by making one of the atoms of oxygen bivalent and the other quadrivalent

$$\begin{matrix} H \\ \\ H \end{matrix}\Large{>}O=O.$$

The valency of an element, defined as atomic weight ÷ equivalent weight, is identical with that obtained from a consideration of its saturation capacity for hydrogen *when there is only one atom of the element in question present in the molecule.* Thus in marsh gas, CH_4, the carbon atom is

quadrivalent when regarded from either point of view, but
where two (or more) atoms of the same kind are directly
united in a molecule, the two definitions lead to divergent
results. For instance, there is only one possible structural
formula for ethane;

$$\begin{array}{c} H \\ H \end{array}\!\!\!>\!C\!-\!C\!<\!\!\!\begin{array}{c} H \\ H \\ H \end{array}$$

where the carbon atoms are each represented as having
four valencies; indeed the development of organic chemistry
has proceeded for upwards of a century on the accepted
quadrivalency of carbon. On the other hand, in ethane
24 grm. of carbon are united with 6 grm. of hydrogen. The
equivalent weight of carbon is therefore 4. Dividing the
atomic weight of carbon, 12, by this equivalent weight,
carbon becomes tervalent; a conclusion quite inconsistent
with the accepted structural formula.

Another and similar instance is to be found in ferric
chloride. At high temperatures the compound has the
molecular formula $FeCl_3$, as has been mentioned. Here the
iron atom is unquestionably tervalent. At lower temperatures the molecular formula is Fe_2Cl_6, and in this case
a graphical formula can only be written with quadrivalent
iron atoms:

$$\begin{array}{c} Cl \\ Cl \\ Cl \end{array}\!\!\!>\!Fe\!-\!Fe\!<\!\!\!\begin{array}{c} Cl \\ Cl \\ Cl \end{array}$$

Ethylene, C_2H_4, presents a somewhat different problem.
Since hydrogen atoms are univalent, they cannot possibly
link two other atoms together. This must be done by the
carbon atoms:

$$\begin{array}{c} H \\ H \end{array}\!\!\!>\!C\!-\!C\!<\!\!\!\begin{array}{c} H \\ H \end{array}$$

Here the carbon atom appears to be tervalent, a difficulty overcome by doubly linking the carbon atoms:

$$\begin{array}{c} H \\ \diagdown \\ H \diagup \end{array} C = C \begin{array}{c} H \\ \diagup \\ \diagdown H \end{array}$$

Such a formula implies that in ethylene there are dormant valencies between the carbon atoms, a view supported by the unsaturated character of this compound. Indeed in those cases in which this double linking is necessary in writing a structural formula, the compound always shews unsaturation in the sense already indicated. The case of acetylene

$$H-C\equiv C-H$$

differs only in degree from that of ethylene. Here the property of unsaturation is more pronounced.

When ammonia combines with hydrogen chloride, the formation of ammonium chloride may be expressed graphically thus:

$$\begin{array}{c} H \\ \diagdown \\ \\ | \\ H \end{array} N \begin{array}{c} H \\ \diagup \\ \end{array} + H-Cl = \begin{array}{c} H \\ | \\ H \diagdown | \diagup H \\ N \\ H \diagup \diagdown Cl \end{array}$$

(van't Hoff)

the nitrogen, tervalent in ammonia, becoming quinquevalent in ammonium chloride. In general terms, any reaction in which substances *combine* (that is, unite to form a single compound) involves the supposition either that dormant valencies are caused to function, or that, in the case of some elements, valency is a variable property. It will have been noted that no information concerning valency can be obtained from compounds of unknown molecular weight.

Ammonium salts can undergo a large number of reactions in which the ammonium radical, NH_4, retains its individuality, passing from one combination to another without disintegration. Thus, when ammonium chloride is treated with concentrated sulphuric acid, hydrogen chloride is evolved and ammonium sulphate is produced:

$$2NH_4Cl + H_2SO_4 = (NH_4)_2SO_4 + 2HCl.$$

Similarly, when solutions of silver nitrate and ammonium chloride are mixed, double decomposition takes place, silver chloride being precipitated and ammonium nitrate remaining in solution:

$$NH_4Cl + AgNO_3 = AgCl + NH_4NO_3.$$

A group of atoms, such as the nitrogen atom united with four hydrogen atoms in the ammonium salts, which is capable of maintaining its integrity throughout a number of chemical changes, is known as a *compound radical*.

A more modern way of expressing the constitution of ammonium chloride (due to Werner) is to write the formula thus:

$$\begin{bmatrix} H & & H \\ & N & \\ H & & H \end{bmatrix} Cl$$

in which expression is given to the individuality of the univalent ammonium group within the square brackets, while the ionic character of the chlorine (Chap. x) is emphasized by writing it outside the brackets.

The ammonium group resembles sodium or potassium in its capability of replacing the hydrogen of an acid, either partly or entirely, as in the sulphate, $(NH_4)_2SO_4$, and the bisulphate, $(NH_4)HSO_4$. The salts it forms also bear a family resemblance to those of sodium and potassium in their properties. The resemblance is close enough to have led chemists in the past to regard ammonium as a metal, and many attempts to isolate such a metal have been made. For instance, when sodium amalgam is placed in a concentrated solution of ammonium chloride, a sponge-like mass of considerable bulk is slowly formed—the so-called ammonium amalgam—the speculation being that this might be an alloy of the 'metal' with mercury. On standing, it evolves ammonia rapidly, and soon leaves nothing but mercury. It is probably mercury aerated into a sponge by ammonia and hydrogen. There is, in fact, no evidence that the group (NH_4) is capable of an independent existence (except as an ion, Chap. x), though there is no *a priori* reason why two such radicals should not unite to form a double ammonium compound, N_2H_8, analogous to the union of two methyl radicals, CH_3, to form ethane, C_2H_6, or of two ethyl radicals, C_2H_5, to form butane, C_4H_{10}. This linking in the case of ammonium groups has never been accomplished. The name ammoni*um* is a relic of the ancient belief in the metallic nature of this radical.

Ammonium salts can generally be prepared by the addition of the required acid to a solution of ammonia, followed by evaporation to the crystallizing point. The crystals are usually anhydrous, few ammonium salts containing water of hydration. For the greater part they are isomorphous with the corresponding salts of sodium and potassium, and like them they are characterized by considerable solubility. Unlike the salts of sodium and potassium, ammonium salts on heating volatilize without fusion, usually dissociating into the free acid and ammonia. The density of the vapour of ammonium chloride for instance is about half that required by the formula NH_4Cl, and this can only be explained by assuming the practically complete dissociation of the salt into hydrogen chloride and ammonia:

$$NH_4Cl \rightleftharpoons NH_3 + HCl.$$

The existence of both these gases in the vapour from ammonium chloride can be demonstrated by the following experiment.

A porous tobacco pipe stem is passed through a glass tube in the manner shewn in Fig. 26. The glass tube contains a little ammonium chloride, which can be vaporized by heat. By means of the indiarubber ball, a gentle stream of air can be forced through the porous pipe into a beaker containing water, coloured with a little red litmus. If the vapour contains both ammonia and hydrogen chloride, the lighter ammonia should diffuse through the porous wall of the tobacco pipe more rapidly than the denser hydrogen chloride, and on being swept into the beaker by the air

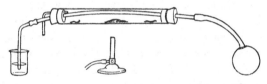

Fig. 26

current, should turn the red litmus blue. The removal of more ammonia than hydrogen chloride from the glass tube would leave a preponderance of the latter gas, and a piece of blue litmus paper placed in the glass tube would become red.

Baker has shewn that, after intensive drying by means of phosphorus pentoxide, ammonia and hydrogen chloride can be mixed without combination taking place, and that ammonium chloride, when exhaustively desiccated, volatilizes without dissociation, giving a vapour density which corresponds very closely to that required by the molecule NH_4Cl.

Nitric Acid and its Derivatives. Since nitrogen and oxygen combine under the influence of electric sparks, it is only to be expected that the atmosphere should contain traces of nitric and nitrous acids after thunderstorms. These acids are rapidly dissolved by rain and find their way into the soil, this being one source of naturally occurring nitrates. The fermentation of nitrogenous organic matter is used as a commercial process for obtaining potassium nitrate in southern countries. The 'nitre plantations,' in which manure and similar organic refuse is exposed to a moist and warm atmosphere, produce potassium nitrate as an efflorescence on the side of the heaps exposed to the sun. This compound is frequently found in caves and underground cellars, as in the catacombs of Paris, and in the wine caves in Champagne and other districts. Large quantities have in the past been imported from India and China. In Chile, in the district on the eastern slope of the Andes, there are extensive deposits of sodium nitrate, the famous caliche beds. These extend for hundreds of miles, having a width of from two to three miles and a varying thickness of from one and a half to twelve feet. Little of this material is at present worked which does not contain more than 17 per cent. of sodium nitrate, the average being about 40 per cent. The very soluble nitrate is dissolved out by water, and the liquors after settling allowed to evaporate under combined solar and artificial heating. The mother liquors which drain from the crystals of sodium nitrate contain considerable quantities of sodium iodate, from which most of the world's supply of iodine is now obtained.

Nitric acid is prepared from nitrates by distillation with concentrated sulphuric acid. On the laboratory scale the operation is carried out in a glass retort, the acid being received in a glass flask (Fig. 27). On the manufacturing scale, iron retorts with earthenware or glass cooling and receiving vessels are used. The mixture of sulphuric acid and nitrate (usually sodium nitrate) is heated. At a moderate temperature the reaction produces sodium bisulphate,

$$NaNO_3 + H_2SO_4 = NaHSO_4 + HNO_3. \quad (1)$$

At higher temperatures the normal sulphate is formed,

$$2NaNO_3 + H_2SO_4 = Na_2SO_4 + 2HNO_3. \quad (2)$$

Equation (2) represents the more economical process from the theoretical point of view, but in practice it is found that the high temperature required leads to the decomposition of a considerable proportion of the nitric acid,

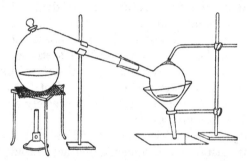

Fig. 27

and a further objection is that the normal sodium sulphate is difficult to fuse, and consequently hard to remove from the retort. In practice, manufacturers use quantities calculated to produce equal weights of the normal and acid sulphates. The decomposition of the acid during distillation is dealt with either by distilling under reduced pressure, or by mixing the gaseous products of decomposition with air and passing them through a water scrubber, where nitric acid is re-formed. The proportion of acid so recovered amounts to about 3 per cent. of the total yield.

In the laboratory preparation, the first distillate of nitric acid can be purified by mixing it with an excess of concentrated sulphuric acid and redistilling at as low a temperature as possible. This product is almost pure, especially if the second distillation be carried out under reduced pressure.

This preparation is an instance of a quite general method of preparing acids from their salts, of which several examples have already been given; the distillation of a salt with a less volatile acid. The acid to be prepared must obviously be volatile, and must also be unaffected by the acid used at the temperature of the distillation. Concentrated sulphuric acid is the less volatile acid usually employed, but other acids, such as phosphoric acid for instance, can be used. It is sometimes erroneously supposed that in such cases the 'strong' acid displaces the 'weak' one. The 'strength' or 'avidity' of the two acids concerned is not here in question. It is solely a matter of the relative vapour pressures.

Although dilute nitric acid, as will be seen later, is by no means unstable, the concentrated acid rapidly decomposes at ordinary temperature, evolving oxides of nitrogen and oxygen, and becoming more dilute. When distilled, the pure acid begins to boil at about 86° C., giving off much gas with but little liquid. The temperature rises during the distillation and becomes steady at 120° C. (at 760 mm. pressure) when the liquid distils without further change, the composition of the acid in the retort being identical with that of the distillate (cf. hydrochloric acid). The specific gravity of the distilled liquid, which contains 68 per cent. of nitric acid, is 1·42. This constant boiling mixture of nitric acid and water is known commercially as *aqua fortis*. *Fuming* nitric acid is the product obtained from the first distillation of nitrates with sulphuric acid.

It is a brownish liquid of specific gravity 1·52; contains 88–90 per cent. of nitric acid, and is the most concentrated form sold commercially. The reddish colour which nitric acid develops is due to the presence of dissolved nitrogen peroxide, one of the products of its decomposition. This may be removed by passing dry air through the liquid. Indeed, if dry air be passed through nitric acid in *any* stage of concentration until equilibrium is reached, the product is always the constant boiling mixture above mentioned.

Nitric acid is usually regarded as a powerful oxidizing agent. This property can be best understood by considering the manner in which the acid decomposes. It has already been seen that the products of decomposition are water, oxygen, and oxides of nitrogen; and also that these substances unite to form nitric acid. The reaction

$$4HNO_3 \rightleftharpoons 2H_2O + 4NO_2 + O_2$$

being reversible, it follows from this that the oxidizing power of nitric acid must depend upon its degree of dilution. For instance, very dilute nitric acid does not liberate iodine from hydriodic acid, nor sulphur from hydrogen sulphide, whereas moderately concentrated acid very readily oxidizes both these compounds. When dropped upon hot charcoal or sawdust, fuming nitric acid instantly causes deflagration. At 300° C. it oxidizes practically all organic compounds into carbon dioxide and water, a reaction much used in ultimate organic analysis. Nitric acid is used for the conversion of ferrous iron into the ferric condition in analytical work. When heated with ferrous sulphate solution in excess, in presence of sulphuric acid, nitric acid oxidizes some of the ferrous sulphate to ferric sulphate, being itself completely reduced to nitric oxide, NO, *but no further*:

$$6FeSO_4 + 3H_2SO_4 + 2HNO_3 = 3Fe_2(SO_4)_3 + 4H_2O + 2NO.$$

This reaction provides a method for obtaining nitric oxide quantitatively from nitric acid, and in a very pure condition. Nitric oxide combines with ferrous sulphate in cold solution to form a dark brown unstable substance of indefinite composition, to which the formula $FeSO_4(NO)_x$ is usually assigned. Use is made of this reaction in what is known as the 'ring' test for nitric acid and nitrates in general. A solution of the substance to be tested is added to excess of a concentrated solution of ferrous sulphate and the mixture well cooled. Concentrated sulphuric acid is cautiously added to this in such a way that the heavy acid for the greater part sinks to the bottom of the test tube, and the dark ring appears where the surface of the sulphuric acid meets with the lighter solution. On warming, the dark compound decomposes with evolution of nitric oxide.

Nitric acid attacks all the better-known metals except gold and platinum, and with the only important exceptions of tin and antimony, takes them into solution in the form of the metallic nitrate. Fuming nitric acid has no action on tin, but more dilute acid reacts violently with it, converting it into the insoluble hydrated oxide $(SnO_2)_x(H_2O)_y$, sometimes termed metastannic acid. Use is made of this in estimating tin in its alloys. A bronze coin, for example, if treated with moderately concentrated nitric acid till no further action takes place, leaves behind all the tin in this hydrated insoluble form. Very dilute and cold nitric acid slowly dissolves tin in the form of the stannous salt. Antimony, too, is converted into the hydrated oxide but not dissolved. Aluminium is only very slowly dissolved. Silver, lead, copper, and mercury are dissolved by nitric acid of almost any concentration; in the case of mercury, excess of the acid yields mercuric nitrate, $Hg(NO_3)_2$, while excess of the metal yields mercurous nitrate, $HgNO_3$.

Iron is easily dissolved by nitric acid over a wide range of dilution, the more concentrated acid producing ferric nitrate, $Fe(NO_3)_3$, while very dilute acid produces ferrous nitrate, $Fe(NO_3)_2$. Iron, when placed in nitric acid of specific gravity 1·45, is said to assume the 'passive' condition. No apparent action takes place, but the iron becomes immune to the attack of more dilute acid. This behaviour is usually attributed to the formation of a thin protective covering of magnetic oxide, Fe_3O_4. Since passivity is destroyed by even gentle rubbing, it must be due to a surface protection of some kind. It is interesting to note that iron vessels are extensively used in operations in which they come into contact with nitric acid, sometimes at high temperatures, and yet are very little corroded. It is found that the presence of sulphuric acid apparently protects the iron from attack, as in the case of the manufacture of nitric acid from nitre, and in the production of organic nitro-compounds in which a mixture of concentrated nitric and sulphuric acids is used.

The *gaseous* products of the interaction between nitric acid and metals differ so much with the conditions that it is difficult to predict in any particular case what these gases will be, and it is very rare to find a single, or even approximately pure, gas evolved.

If the assumption be made that the first action of the metal on the acid is to displace hydrogen (an assumption which has some basis, since hydrogen has in fact been collected from the action of both magnesium and iron on dilute and cold nitric acid) reduction products of the acid would be expected. The most important of these may be summarized in the order of the stages of reduction; nitric peroxide, (nitrous anhydride ?), nitric oxide, nitrous oxide, nitrogen, ammonia. The first stage of the action of a metal on nitric acid may be regarded as the production of the metallic nitrate with displacement of hydrogen. The second stage, the reducing action of the hydrogen so displaced on the remaining nitric acid, with production of nitrogen peroxide. Whether the next stage is to be a further reduction of the peroxide, or the reduction of more nitric acid, would depend on the relative stabilities of the two compounds. Since, as has been shewn, the stability of nitric acid increases with dilution, the more dilute the acid the greater its resistance to reduction, and the greater the probability that the peroxide will be the compound to be attacked. This principle should hold at all stages of reduction, and it follows that the more dilute the acid the lower down the scale (i.e. towards ammonia) will the products tend, other conditions being the same. Another factor in determining the nature of the gaseous products is temperature. Since rise of temperature tends to decrease the stability of nitric acid, thus rendering it more easy to reduce, heating therefore would make it more likely that the acid itself would be reduced rather than the less oxidized gases, while cooling would tend to a production of the more highly reduced gases. A third factor in determining the gaseous products formed is the nature of the metal itself. Metals vary greatly in what is termed their electro-chemical character. This finds qualitative expression in the ease or difficulty with which they displace hydrogen from acids. Thus, magnesium is more electro-positive than zinc, and the latter than copper, or mercury, or silver (see Chap. XI). As the reducing power of hydrogen at the moment of its liberation from an acid depends upon its potential, the more electro-positive the metal the greater the reducing power of the hydrogen it displaces, and therefore the more complete will the reduction tend to become.

These tendencies (and it should be clearly understood

that *tendencies only* have been here considered) can be summarized in the following diagram:

Dilute acid; low temperature; electro + metal.

$$\text{HNO}_3. \quad \text{NO}_2. \quad (\text{N}_2\text{O}_3). \quad \text{NO}. \quad \text{N}_2\text{O}. \quad \text{N}_2. \quad \text{NH}_3. \quad \text{H}_2.$$

Concentrated acid; high temperature; less electro + metal.

The reactions of nitric acid, especially with strongly electro-positive metals, are influenced and complicated by the presence or absence of traces of nitrous acid, but these need not be considered here.

The composition of the gaseous products of the action of nitric acid on metals depends therefore on:

1. The metal used.

2. The concentration of the acid.

3. The temperature at which the action takes place.

4. The presence of other substances, including the products of the reaction itself.

It may be convenient to consider a few specific cases. When copper and nitric acid are brought together under ordinary laboratory conditions, the gases evolved are *mainly* nitrogen peroxide and nitric oxide; the more concentrated the acid the larger being the proportion of peroxide, in accordance with the principles enumerated above. The equations representing the reactions can be deduced thus:

1. The reduction of nitric acid by hydrogen to nitric peroxide.

$$2\text{HNO}_3 + \text{H}_2 = 2\text{H}_2\text{O} + 2\text{NO}_2.$$

2. The *theoretical* liberation of the required amount of hydrogen from the acid by the metal.

$$\text{Cu} + 2\text{HNO}_3 = \text{Cu(NO}_3)_2 + \text{H}_2.$$

On adding equations 1 and 2 (and removing those substances which occur in equal quantities on both sides of the equations)

$$Cu + 4HNO_3 = Cu(NO_3)_2 + 2H_2O + 2NO_2.$$

Or, for nitric oxide:

$$2HNO_3 + 3H_2 = 4H_2O + 2NO.$$

$$3Cu + 6HNO_3 = 3Cu(NO_3)_2 + 3H_2.$$

And on adding:

$$3Cu + 8HNO_3 = 3Cu(NO_3)_2 + 4H_2O + 2NO.$$

Using zinc and fairly dilute acid, at not too high a temperature, nitrous oxide (impure) is produced:

$$2HNO_3 + 4H_2 = 5H_2O + N_2O.$$

$$4Zn + 8HNO_3 = 4Zn(NO_3)_2 + 4H_2.$$

By addition:

$$4Zn + 10HNO_3 = 4Zn(NO_3)_2 + 5H_2O + N_2O.$$

With more dilute acid, and at a low temperature, some ammonia is produced: This is not, of course, evolved *quâ* ammonia, but is converted into ammonium nitrate as it is formed, by the excess of acid present, from which it can be recovered and identified later by adding excess of alkali to the residual solution and gently heating. The equations for its formation are:

$$HNO_3 + 4H_2 = 3H_2O + NH_3.$$

$$4Zn + 8HNO_3 = 4Zn(NO_3)_2 + 4H_2.$$

Adding the equations:

$$4Zn + 9HNO_3 = 4Zn(NO_3)_2 + 3H_2O + NH_3.$$

Metallic nitrates are characterized *as a class* by their great solubility in water, and hence their preparation is easily made by adding the basic oxide or carbonate of the metal to nitric acid, preferably somewhat diluted, the nitrate being separated by crystallization.

The effect of heat on solid nitrates may be summarized as follows:

1. Ammonium nitrate decomposes into nitrous oxide and steam:

$$NH_4NO_3 = 2H_2O + N_2O.$$

This is an instance of decomposition, since the action is not reversible. Nitrous oxide is usually prepared in this way.

2. The nitrates of potassium and sodium fuse below a red heat, and at a somewhat higher temperature evolve oxygen, leaving the metallic nitrite:

$$2KNO_3 = 2KNO_2 + O_2.$$

3. Other metallic nitrates, when anhydrous, for the greater part yield the basic oxide of the metal, and evolve both oxygen and nitric peroxide. The latter gas is usually prepared by heating lead nitrate:

$$2Pb(NO_3)_2 = 2PbO + 4NO_2 + O_2.$$

This may be regarded as a somewhat general method for the preparation of basic oxides. Copper and mercury, for instance, are most conveniently, rapidly, and completely converted into their higher basic oxides by dissolving the metal in excess of nitric acid, evaporating the solution to dryness, and igniting the residue.

The metallic nitrates are more stable than the acid itself, though, being rich in oxygen, they may in general be classified as oxidizing agents. A piece of sulphur, for instance, on being dropped on molten potassium nitrate burns with a brilliant flame. Blotting paper or string soaked in a concentrated solution of potassium nitrate and dried, makes a kind of fuse which on the application of a spark smoulders with deflagration. A spark applied to gunpowder (an intimate mixture of carbon, sulphur, and potassium nitrate), starts a process of oxidation of the two elements which is very rapid and highly exothermic, generating large volumes of hot gases, hence the explosion. Although ammonium nitrate can safely be fused and heated to decomposition, it can be made to explode with terrific violence by means of a suitable detonator. It is, in fact, a high explosive, and most substances of this class are either nitrates, or compounds closely related to nitrates. The molecular structure of nitric acid may be represented by the graphical formula:

$$H-O-N\begin{matrix} O \\ O \end{matrix}$$

nitrates being derived from it by replacement of the hydrogen atom by a metal or a radical:

$$K-O-N\begin{matrix} O \\ O \end{matrix}, \text{ and } R-O-N\begin{matrix} O \\ O \end{matrix}$$

When nitric acid acts upon glycerine, the nitrate of an organic radical, glyceryl, is formed. This is glyceryl trinitrate, sometimes erroneously called nitroglycerine:

$$C_3H_5 \left[-O-N\begin{matrix} O \\ O \end{matrix} \right]_3$$

and is one of the highest explosives known. It is a thickish colourless liquid which, when diluted by absorption in a porous kind of earth known as kieselguhr, forms dynamite. A small drop of the liquid placed on an anvil and struck with a hammer produces a deafening report. When ignited in small quantities, it burns quietly and without explosion. Tri-nitro-cellulose (gun cotton) is also a nitrate of this class.

True nitro-compounds are formed by replacement of the *hydroxyl radical* of nitric acid by an organic group:

$$(HO)-N{\Large\diagup}^{O}_{\diagdown O} \longrightarrow R-N{\Large\diagup}^{O}_{\diagdown O}$$

Nitro-benzene, $C_6H_5NO_2$, and tri-nitro-toluene (T.N.T.) are compounds belonging to this class.

It will be seen that the essential difference between nitrates and nitro-compounds lies in the relationship of the radical to the nitrogen atom. In the nitrates the two are joined *indirectly*, through an atom of oxygen, whereas in nitro-compounds the radical is *directly* attached to the nitrogen atom. (See p. 177.)

Nitrous Oxide. The gas obtained by heating ammonium nitrate, being somewhat soluble in cold water, is usually collected over hot water. If required in dry condition, the steam accompanying the gas can be for the greater part removed by cooling, after which it may be dried by any of the usual desiccating agents and collected over mercury. Traces of the higher oxides of nitrogen can be eliminated by a slow passage of the gas through a concentrated solution of ferrous sulphate before drying. The only probable remaining impurity, nitrogen, can be separated by liquefaction, nitrous oxide being much easier to condense than nitrogen.

Since nitrous oxide, like all the oxides of nitrogen, is endothermic, when combustible substances burn in it they evolve more heat than when burnt in pure oxygen. The gas can be distinguished from oxygen, obviously not by deflagration experiments, but by its greater solubility in cold water, by its indifference to alkaline pyrogallol, and by the introduction into it of a few bubbles of nitric oxide, since, unlike oxygen, it does not yield the reddish nitrogen peroxide, NO_2, with nitric oxide. When inhaled for a short time nitrous oxide produces unconsciousness, and on this account is used as an anaesthetic, in dentistry and in some minor surgical operations. For this purpose the gas should be pure, or at least free from other oxides of nitrogen, since the latter are all toxic. Nitrous oxide is sometimes described as a 'neutral' oxide, in the sense that it shews neither acidic nor basic tendencies. Like carbon monoxide, it can only be brought into combination with water by indirect methods.

The molecular formula is deduced from (a) its vapour density, which is found by direct weighing to be 22, and (b) by heating an easily oxidizable substance such as tin, potassium, iron, or phosphorus in a measured volume of the gas. These remove all oxygen, and the residual nitrogen is found to be equal in volume to the gas taken. From (b) the formula N_2O_x is derived. The value of x is shewn to be unity by subtracting from the molecular weight (44) the weight of nitrogen present (28), the remainder (16) being one atomic weight of oxygen.

Nitric Oxide. As has been shewn, this gas is not produced in pure condition by the interaction of nitric acid with metals, though for experimental purposes it is usually prepared by acting upon copper or mercury with moderately dilute acid, the yield being rather small. This preparation can be purified by absorption in ferrous sulphate and subsequent boiling out of the nitric oxide, since all oxides of nitrogen *higher* than nitric oxide are reduced to the latter by ferrous sulphate, and all oxides *lower* than nitric oxide are neither dissolved nor chemically changed. The pure gas is most conveniently prepared by reducing nitric acid with ferrous sulphate. This is carried out in practice by heating sodium or potassium nitrate with a large excess of a concentrated solution of ferrous sulphate, together with sulphuric acid. The gas can be collected over water, in which it is almost insoluble, or dried by any of the ordinary desiccating agents, and collected over mercury. Combustible substances do not burn readily in nitric oxide unless introduced into the gas at a high temperature. A mixture of the vapour of carbon bisulphide and nitric oxide, on ignition, flashes silently with an intensely actinic blue flame. The conditions under which nitric oxide unites with oxygen to form the peroxide have already been fully described. Cavendish made use of this reaction in one of the earliest estimations of the percentage of oxygen in the atmosphere. Nitric oxide was admitted slowly and cautiously to a measured volume of air so long as any diminution in volume took place, the peroxide being absorbed as it was formed by an alkaline solution. In this way the whole of the oxygen was removed from the volume of air taken.

The molecular formula of nitric oxide is established in manner similar to that for nitrous oxide. Oxidizable substances heated in it remove the oxygen, the volume of nitrogen remaining in this case being one half of that of the original gas. Since one molecule of nitrogen is obtainable from two molecules of the gas, the formula must be NO_x. As the vapour density of nitric oxide is 15 it follows that $x = 1$.

Nitrogen Peroxide is generally prepared by passing the products obtained by heating carefully-dried lead nitrate through a freezing mixture of ice and salt. A brown liquid separates, the oxygen passing through uncondensed.

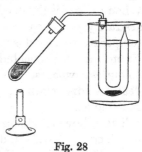

Fig. 28

Gay-Lussac shewed that the gas can be prepared from

two volumes of nitric oxide and one volume of oxygen, the molecular formula being therefore $(NO_2)_n$. Since measurements of vapour density are unreliable when made at temperatures too nearly approaching the boiling point, no trustworthy information concerning the molecular composition of nitrogen peroxide can be obtained below about 30° C. At this temperature the density corresponds neither to the formula NO_2 nor N_2O_4, but has an intermediate value. Nor is the density constant over even a small range of temperature in this region. As the temperature rises the density gradually falls, becoming constant over the range 150° C.–200° C. approximately. Above this, dissociation into nitric oxide and oxygen takes place.

These abnormal densities indicate the existence of two kinds of molecule in equilibrium, the variation representing the extent of the reversible change

$$N_2O_4 \rightleftarrows 2NO_2.$$

Above 150° C. the gas consists entirely of the smaller molecules, but at no temperature does it consist solely of N_2O_4 molecules. Furthermore, on heating the gas from laboratory temperature to 150° C., it gradually deepens in colour, at the latter temperature being almost opaque. When the gas is liquefied the brownish colour remains, but gradually disappears on further cooling, until at lower temperatures a solid is produced which is almost colourless. From this it is concluded that the larger molecules are colourless, whereas the smaller ones are brown, and the degree of dissociation is roughly indicated by the depth of the colour.

The following values for the density of the gas were obtained by Deville and Troost, using dry air as the standard. The density of N_2O_4 is 3·18 in terms of air as unity, and that of NO_2 is 1·59. The percentage number of N_2O_4

molecules which have dissociated is given in the third line below.

Temperature	40° C.	70° C.	90° C.	121° C.
Density	2·46	1·92	1·72	1·62
Percentage dissociation	29·2	65·6	84·8	96·2

The following is an example of a method of making the above calculations from first principles:

Suppose 100 molecules of N_2O_4 be taken, of which x molecules dissociate, each dissociating molecule producing two molecules of NO_2.

There are present therefore $100 - x$ molecules of N_2O_4, each of weight 3·18, and $2x$ molecules of NO_2, each of weight 1·59.

Total weight $= 3·18 (100 - x) + 1·59 \times 2x$.

But as the observed density is 1·92, there are $100 + x$ molecules of *average* weight 1·92. The total weight is therefore $1·92 (100 + x)$.

Whence $1·92 (100 + x) = 3·18 (100 - x) + 1·59 \times 2x$.

and $\qquad\qquad x = 65·6$ (at 70° C.).

In the case of nitrogen peroxide increase of pressure diminishes the degree of dissociation. Such is always the case when the number of molecules (and therefore the volume, or the pressure) is increased by dissociation. When, as in the case of hydrogen iodide, the number of molecules is unaltered upon dissociation, change of pressure is without influence. Such results would be expected from a consideration of Le Chatelier's theorem.

Potassium nitrite is obtained by heating potassium nitrate to a dull red heat; or better, by heating the nitrate with twice its weight of lead:

$$Pb + KNO_3 = PbO + KNO_2.$$

The nitrite is dissolved out by water from the lead oxide, and the solution evaporated. As thus prepared, potassium nitrite is a yellowish rather soft solid, very soluble in water, to which it imparts an alkaline reaction, due to limited hydrolysis. On mixing solutions of the nitrite and silver nitrate, a yellowish precipitate of silver nitrite is formed which can be crystallized from hot water.

When nitrites are treated with dilute acids, nitrous acid is liberated. This compound is very unstable, the aqueous solution having a blue colour somewhat resembling that of dilute cupric salts, which disappears after a time, the liquid evolving nitric oxide while nitric acid remains in solution:

$$3HNO_2 = H_2O + 2NO + HNO_3.$$

The acid has both oxidizing and reducing properties. Substances rich in oxygen, such as permanganates and bichromates, oxidize it to nitric acid, the permanganates being decolorized in the process, and the bichromates turning green. Reducing agents, such as hydriodic acid or hydrogen sulphide, are oxidized by nitrous acid with liberation of iodine and sulphur respectively. Nitrites are easily distinguished from nitrates by these reactions, and also by the evolution of a coloured gas, nitric peroxide, on adding even dilute acids. It is difficult to test for nitrates in the presence of nitrites, but the latter can be easily and completely removed even in the presence of nitric acid by heating the solution containing them with a dilute acid (usually sulphuric) and urea, $CO(NH_2)_2$:

$$CO(NH_2)_2 + 2HNO_2 = CO_2 + 3H_2O + 2N_2.$$

Nitrous acid is commonly used, especially in organic chemistry, for replacing an amino group, (NH_2), by a hydroxyl group, (OH), and this is what actually takes place with urea, the carbonic acid, $CO(OH)_2$, first formed at once breaking down to carbon dioxide and water.

Nitrites are often found in contaminated water, and freely in sewage, where their presence has a special and important significance as indicating putrefaction. A test which shews the presence of the merest traces of nitrites may be illustrated thus: Two beakers, each containing about 250 c.c. of distilled water acidified with dilute sulphuric acid are taken. To one of them a single drop of a 10 per cent. solution of a nitrite is added, followed by the addition of about 20 c.c. of a saturated aqueous solution of creosote to both. The beaker containing the nitrite develops a deep orange colour in a few moments, though the nitrite content is less than one part in 100,000. Carried out in this comparative way, the test will indicate the presence of one part of nitrite in one million parts of solution, the test being unaffected by the presence of nitrates.

From a comparison of the formulae KNO_2 and $(C_6H_5)NO_2$, it might be supposed that nitro-compounds are salts of nitrous acid, but such is not the case. Two compounds whose molecular formulae are both $(C_2H_5)NO_2$ are known, which differ widely in properties and reactions. They are both liquids, but their boiling points are nearly a hundred degrees apart. One of these is a nitrite, and the other a true nitro-compound.

The structural formula for nitrous acid is usually represented $H-O-N = O$, and since nitrites are derived from the acid by replacement of the hydrogen atom by a metal or radical, the structures of potassium nitrite and of ethyl nitrite must be $K-O-N = O$ and $(C_2H_5)-O-N = O$ respectively, whereas, as has been shewn, in the true nitro-compounds the radical is united to the nitrogen atom directly, thus:

$$(C_2H_5)-N{\Large\langle}{}^O_O \quad \text{(nitro-ethane).}$$

Compounds having the same composition and the same molecular weight, such as nitro-ethane and ethyl nitrite, are said to be *isomeric*.

When nitric oxide and nitrogen peroxide are mixed, the change of volume which takes place is very slight. Apparently the gases do not unite to form a compound, N_2O_3, under ordinary laboratory conditions, or at most the compound can only be formed in mere traces. When mixed in equivalent proportions (and such a mixture can be prepared by warming moderately concentrated nitric acid with white arsenic, As_2O_3) and passed through a tube cooled by ice and salt, a greenish liquid results. This, on further cooling, yields greenish blue crystals corresponding in composition to the formula $(N_2O_3)_x$. On volatilization, these yield a gas whose density is considerably less than is required by the formula N_2O_3. It is generally held that combination between nitric oxide and nitric peroxide does take place, to a minute extent, in the gaseous phase under ordinary conditions, and to a greater extent in the liquid and solid phases, the compound being regarded as the anhydride of nitrous acid.

The anhydride of nitric acid is said to have been obtained by withdrawal of the elements of water from nitric acid by means of phosphorus pentoxide. The compound is only of theoretical interest, and owing to its unstable nature nothing is known of its molecular constitution, which may be represented as $(N_2O_5)_x$.

Cyanogen. When the electric arc passes between carbon poles in an atmosphere of nitrogen, small quantities of cyanogen, C_2N_2, are formed. The compound is most conveniently prepared by heating mercuric cyanide, $Hg(CN)_2$, which decomposes thus:

$$Hg(CN)_2 = Hg + C_2N_2.$$

The gas, which should be collected over mercury since it hydrolyses with water, is poisonous and burns with a peach-blossom coloured flame, producing carbon dioxide and nitrogen. In some of its reactions the gas resembles chlorine; for instance, when passed into an aqueous solution of potash, cyanide and cyanate of potassium are formed, together with other compounds:

$$2KOH + C_2N_2 = KCN + KCNO + H_2O.$$

The gaseous analysis of cyanogen affords an instance of a method of determining the molecular composition of a compound as the result of a single experiment:

8·3 c.c. of the gas were mixed over mercury with 19·7 c.c. of oxygen and exploded by an electric spark, when 28 c.c. of gaseous products were obtained. Of these, 16·6 c.c. were absorbed on the addition of caustic potash, and a further absorption of 3·1 c.c. followed on the addition of pyrogallol, the residue being nitrogen. Hence 8·3 c.c. of the gas required 16·6 c.c. of oxygen, yielding 16·6 c.c. of carbon dioxide and 8·3 c.c. of nitrogen. That is, 1 volume of cyanogen reacts with 2 volumes of oxygen to produce 2 volumes of carbon dioxide and 1 volume of nitrogen. From Avogadro's theorem it follows that the only equation which is consistent with these numbers is:

$$C_2N_2 + 2O_2 = 2CO_2 + N_2.$$

The compound is not here of much importance, though its derivatives are both numerous and important.

CHAPTER VI

SULPHUR AND ITS COMPOUNDS

SULPHUR occurs freely in the uncombined state in most volcanic districts, large quantities being found in Sicily. Most metals are found in combination with sulphur as sulphides or sulphates, e.g. iron pyrites, FeS_2; galena, PbS; blende, ZnS; gypsum, $CaSO_4$; celestine, $SrSO_4$; and barytes, $BaSO_4$. It also occurs in small quantity in certain organic compounds, as albumin, and in a few essential oils such as mustard oil and oil of garlic.

From the commercial standpoint native sulphur is the most important. The earthy deposits in which the sulphur content is greater than 12–15 per cent. are heated out of contact with the air until the sulphur melts and runs to the lower part of the vessel, from which it is recovered on cooling in large cakes. The poorer ores are heated in closed vessels to a temperature above the boiling point, when the sulphur distils off and is condensed in water-cooled receptacles. Further purification is effected by resublimation, the vapour passing into air-cooled stoneware chambers where it is at first cooled so suddenly as to pass from the gaseous to the solid phase without liquefying. Such sulphur is in a very finely divided condition, known technically as 'flour' or 'flowers' of sulphur. As the temperature of the cooling chamber rises the sulphur melts, and is then run off into wooden moulds where it solidifies in the cylindrical masses sold as brimstone. Much sulphur of a very pure kind is recovered from the sulphuric acid used in alkali manufacture and other works. Combined sulphur, such as that in iron pyrites, is rarely extracted in the elementary state. These sources are used for supplying the large quantities of sulphur dioxide required in the manufacture of sulphuric acid.

If a few grammes of sulphur be heated in a test tube, the solid first melts to a clear amber-coloured somewhat mobile liquid. As the temperature rises, it darkens and becomes more and more viscous, until the test tube may be inverted without the now nearly black material shewing much tendency to flow. At a still higher temperature some of the mobility returns, though the colour remains dark, and at 444° C. it boils. On cooling, these changes take place in the reverse order, but since sulphur is a very bad conductor of heat, the contents of the tube should be kept well agitated, to equalize the temperature throughout the mass.

Sulphur exists in many allotropic modifications of which only the more important will be considered. Fusing sulphur of any kind and allowing it to resolidify produces the prismatic or monoclinic variety. This is shewn as a laboratory experiment by heating brimstone in a crucible until it is completely liquefied, and then withdrawing the flame. The crucible should not be too small and should be nearly filled by the molten mass. When a crust of solid sulphur has completely covered the surface, two holes are bored through it and the still liquid portions are then poured out through one of them, the other serving for the admission of air. When no more liquid drains away the remaining mass is allowed to cool, after which the surface crust can be removed. The interior shews long needle-like transparent crystals of a honey colour. On standing for a few days these become lighter in colour, opaque, and more brittle. Crystals of the same kind can be seen on examining the centre of a broken roll of brimstone with a lens.

The prismatic modification is unstable at all temperatures below 96° C., and soon passes over into the rhombic

variety, hence the opacity of the crystals, which however
maintain their general shape. The needles are now built
up of very small octahedra, not in the least needle-like in
form. They have become
masses, crystalline in form,
but constituted of crystals
belonging to a different
system. The two kinds of
crystals are illustrated in
Fig. 29. This rhombic
variety is stable at all
temperatures below 96° C.
It is not unusual to find
near the soffioni or hot
sulphur springs of Sicily,

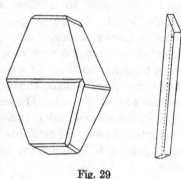

Fig. 29

tiny crystals of prismatic sulphur quite close to the hot
vent, while a little further away the sulphur crystals are of
the rhombic variety, each form being stable in its own en-
vironment.

Prismatic and rhombic sulphur provide an instance of
the equilibrium which sometimes exists between allotropes.
These two forms are both stable at 96° C. Above this temperature only the prismatic form is stable, and below it
only the rhombic; the one passing into the other reversibly
as the temperature rises above or falls below 96° C., which
is defined as the transition temperature. This reversible
process is known as *enantiotropy*, in contradistinction to
the irreversible phenomenon, *monotropy*, shewn by the two
allotropes of phosphorus (Chap. VII).

A transition point in some respects resembles a melting
point, as indicating the temperature at which two phases
can coexist in equilibrium, but with the difference that the
temperature cannot be raised above the melting point
without change of state, whereas the passage of a substance

from the metastable to the stable modification is often somewhat slow. Rhombic sulphur, for instance, may be heated for some time above 96° C. without wholly passing into the prismatic variety, though it does so ultimately.

Rhombic sulphur melts at 115° C. In order to verify this the powdered material must be rapidly heated to this temperature. If slowly heated, it has time to pass over, partly or entirely, into the prismatic variety which melts at 120° C., and thus the true melting point of rhombic sulphur is obscured. It is characteristic of enantiotropic substances that the melting points of *both* forms are *above* the transition temperature. Both these forms of sulphur dissolve in carbon bisulphide, the solution on evaporation depositing clear amber-like crystals of the rhombic variety.

When sulphur is heated to near its boiling point and rapidly cooled, a third form, an india rubber-like mass of plastic sulphur is obtained. The sudden cooling required may be brought about by pouring the hot liquid in a thin stream into cold water. This modification is unstable and soon passes over into the rhombic variety, the transition being accelerated by mechanical agitation. Plastic sulphur does not dissolve in carbon bisulphide.

On extracting flowers of sulphur exhaustively with carbon bisulphide, an insoluble residue remains which obviously does not belong to any of the three classes hitherto mentioned. This form is sometimes described as amorphous sulphur, though, as the word signifies non-crystalline, it might quite properly be applied to the plastic variety. Amorphous sulphur is a vague description. The element may be obtained by precipitation in a number of ways, as, for instance, by acidifying yellow ammonium sulphide, and often these precipitated forms of sulphur, which are generally almost white, are insoluble in carbon bisulphide. It is probable that there are several varieties of this type but they are not here important. They can all be made to pass over into one or other of the crystalline allotropes.

The atomicity of the molecule of gaseous sulphur varies with the temperature. At about 500° C. the vapour density corresponds to a hexatomic molecule (as a mean value) but gradually falls as the temperature rises. From 900° C. to 1700° C. the density is nearly constant, and corresponds to a diatomic molecule. Above 1700° C. the density again begins to fall, indicating a dissociation into single atoms.

Cryoscopic and ebullioscopic methods (Chap. x) indicate that the molecule of sulphur when in solution consists of *eight* atoms.

All the forms of sulphur are practically insoluble in water, though the crystalline varieties dissolve to a slight extent in certain oils. When boiled with caustic alkalis, sulphur passes into solution in the form of the metallic sulphide. Heated with concentrated sulphuric or nitric acid it is oxidized, by the former to sulphur dioxide and by the latter to sulphuric acid. Molten sulphur combines readily with many metals, as may be seen by heating sulphur with filings of iron, zinc, or copper.

When hydrogen is passed through boiling sulphur, combination takes place with the formation of hydrogen sulphide, H_2S. This compound is usually prepared by the action of dilute sulphuric or hydrochloric acid on ferrous sulphide, though most metallic sulphides yield the gas when warmed with dilute acids, this reaction being used as a preliminary qualitative test for sulphides in general.

$$FeS + H_2SO_4 = FeSO_4 + H_2S.$$

The gas may be collected over hot water with but slight loss, or it may be collected over mercury if quite dry,

otherwise it reacts with the mercury and tarnishes it.
Anhydrous calcium chloride is generally used as the desiccating agent, though there is a slight liberation of hydrogen
chloride. For more careful work, phosphorus pentoxide
should be employed. Concentrated sulphuric acid cannot
be used, as it oxidizes hydrogen sulphide with precipitation
of sulphur:

$$H_2SO_4 + H_2S = 2H_2O + SO_2 + S.$$

Since ferrous sulphide usually contains traces of uncombined iron, the gas prepared from it is liable to be contaminated with free hydrogen. A purer product, free from
hydrogen, may be obtained by heating the sulphide of
antimony with concentrated hydrochloric acid:

$$Sb_2S_3 + 6HCl = 2SbCl_3 + 3H_2S,$$

since in the first place the uncombined metal is not generally found in the native sulphide, stibnite, and secondly
hydrochloric acid does not attack the metal.

Hydrogen sulphide is found as a bye-product in many
manufactures. It occurs in raw coal gas to a considerable
extent, and is removed by passing the gas firstly over
layers of quicklime or slaked lime:

$$CaO + 2H_2S = CaH_2S_2 + H_2O,$$

and finally, to remove the last traces of hydrogen sulphide,
over a porous kind of ferric oxide known as bog iron ore:

$$Fe_2O_3 + 3H_2S = Fe_2S_3 + 3H_2O.$$

The evil-smelling waste product from the first purification
is mainly calcium hydrogen sulphide, and finds some
application in agriculture under the name of gas lime, or
spent lime. That from the second is exposed to the air,
where, by the action of oxygen, the ferric oxide is re-formed
and can be used again:

$$2Fe_2S_3 + 3O_2 = 2Fe_2O_3 + 6S.$$

This process may be repeated many times, but as the free sulphur accumulates the oxide becomes less and less active, and finally it is passed to the sulphuric acid works where the sulphur is recovered as the dioxide.

Hydrogen sulphide is found in traces in volcanic gases, and in decaying organic matter which contains sulphur. The odour is very pronounced in putrefying eggs. The 'hepatic' springs of Harrogate and Baden, which have long been credited with curative properties for liver disorders, contain the gas. It has a burning taste and when breathed even in small quantity produces depression and faintness. The coefficient of solubility of hydrogen sulphide in water at 0° C. is 4·37, and at laboratory temperature about 3. The solution, which possesses feebly acidic properties, is known as hydrosulphuric acid. The gas may be completely removed from solution by boiling. Since hydrogen sulphide is easily oxidized by dissolved oxygen with liberation of free sulphur, an aqueous solution, unless made with well boiled out water and kept from contact with air, is not permanent. For this reason, and also because of its low solubility, aqueous solutions are little used in analytical operations. A more concentrated and more permanent solution can be made by using a mixture of glycerine and water.

In a free supply of air, the gas burns with a blue flame yielding sulphur dioxide and water, and as a waste product it is usually disposed of in this way, the dioxide being converted into sulphuric acid. In a limited supply of air, the hydrogen burns to water and the sulphur is deposited. This can be seen by setting fire to a jar of the gas, when the walls of the vessel become covered with the liberated sulphur.

Many metals are converted into the sulphide by contact with the gas, hydrogen being liberated. Mercury and silver are tarnished, very readily if moisture be present. The effect of contact with cooked eggs on silver spoons and dishes is well known. The beautiful patina of silver sulphide on silver articles, known to jewellers as 'gun' metal, or 'oxidized' silver, may be produced by exposing silver articles to an atmosphere of hydrogen sulphide, or better, by gently warming the cleaned silver in a dilute solution of ammonium sulphide, afterwards washing with water and polishing with chamois leather. Since silver sulphide is dissolved by potassium cyanide, a solution of the latter may be used for cleaning silver stained by sulphide. The stain produced on a silver coin is sometimes used as a test for the presence of hydrogen sulphide. A more sensitive one is a drop of a solution of a lead salt, preferably the acetate, on a piece of filter paper, which gives a brown colour when it comes in contact with traces of the gas. This is the test generally used in gas works. Another is to expose paper moistened with a drop of caustic soda to the gas to be tested. If hydrogen sulphide be present it will react with the soda to form sodium sulphide, and this, like all the soluble sulphides, gives a purple colour with a drop of sodium nitro-prusside solution.

Since the basis of most oil colours is white lead, the basic carbonate, oil paintings exposed to an atmosphere containing even minute traces of hydrogen sulphide become dark in course of time from the slow production of lead sulphide. All processes of 'restoring' such pictures turn upon the conversion of the sulphide into colourless lead sulphate by some method of oxidation, whether this be by exposing the picture to light and air, by washing with a solution of hydrogen peroxide, by spraying with ozonized air, or by other means (see p. 83).

Metallic sulphides are not, in general, converted into sulphates quite so readily as this, though precipitated copper sulphide, CuS, is oxidized to the sulphate quite appreciably, if washed with water which contains dissolved oxygen.

The heat of formation of hydrogen sulphide being small (2·7 K. per gramme-molecule in the gaseous condition, and 7·3 K. in solution) not much energy is required to decompose it into its elements. It is therefore a very convenient reducing agent, and since when it acts in this way sulphur is liberated, there is a visible indication of its reducing action. The sulphur is furthermore very easily removable by filtration. For instance, when passed through a solution of ferric chloride, the latter is reduced to the ferrous condition:

$$2FeCl_3 + H_2S = 2FeCl_2 + 2HCl + S.$$

Acidified solutions of permanganates are decolorized by it, and those of bichromates become green. The halogens react with hydrogen sulphide to produce the halogen acids:

$$H_2S + Cl_2 = 2HCl + S.$$

With sulphur dioxide it reacts to form water:

$$SO_2 + 2H_2S = 2H_2O + 3S,$$

sulphur being precipitated in all these cases. It is oxidized by nitric acid in all but very dilute solutions, hence the use of nitric acid in analytical processes for the removal of hydrogen sulphide. Since the metallic sulphides are salts of hydrosulphuric acid, it should be possible to prepare them by the general reaction of the acid on the hydroxide or carbonate of the metal. A more convenient method, however, is found in either the direct union of sulphur with the metal itself, or in the precipitation of the sulphide from a solution of the metallic salt. In those cases where the metallic sulphide is unacted on by dilute acids, it is sufficient to pass hydrogen sulphide through an aqueous solution of one of the salts of the metal. For example, the whole of the copper may be precipitated in the form of the sulphide from a solution of copper sulphate by passing hydrogen sulphide through it till the blue colour has disappeared:

$$CuSO_4 + H_2S = CuS + H_2SO_4.$$

The sulphide of tin may be obtained in a similar way, by passing the gas through a *dilute* solution of stannous chloride. Since this reaction is reversible:

$$SnCl_2 + H_2S \rightleftharpoons SnS + 2HCl,$$

if a *concentrated* solution of the salt be used, the whole of the tin will not be precipitated, since the concentration of the hydrochloric acid formed by the reaction will reach the point at which equilibrium is established, and the reaction thereupon ceases. If it is desired to precipitate the whole of the tin, the solution should be filtered from the first precipitate, diluted, and treated with more gas, and this operation must be repeated until further dilution and passage of hydrogen sulphide produces no more precipitate. This is often an important point in analytical operations. In those cases where the required sulphide reacts with dilute acids, it may be precipitated from a solution of the salt by the addition of a soluble sulphide, ammonium or sodium sulphide being generally used:

$$(NH_4)_2S + ZnSO_4 = (NH_4)_2SO_4 + ZnS.$$

If a small volume of dry hydrogen sulphide be collected over mercury in a sparking tube and a series of electric sparks passed through it, the tube becomes coated with sulphur, but on regaining the temperature of the room no change in the volume of gas is observed. A piece of tin or cadmium heated in the gas removes all the sulphur, but here again the volume of gas is not changed. Hydrogen sulphide, then, contains its own volume of hydrogen; that is, one molecule of it produces one molecule of hydrogen, and therefore the molecular formula must be H_2S_x. Since the density is 17 times that of hydrogen, the molecular weight must be 34. As the atomic weight of sulphur is 32, it is obvious that there can be but one atom of sulphur in the molecule of hydrogen sulphide. (Cf. nitrous and nitric oxides.)

The laboratory reagent ammonium sulphide is made by saturating a solution of ammonia with hydrogen sulphide. The colourless liquid so obtained probably contains both the normal sulphide, $(NH_4)_2S$, and the acid sulphide, $(NH_4)HS$. It is easily oxidized by air with liberation of sulphur, as may be seen by pouring a little on a watch glass and leaving it exposed. The incrustation round the neck of the containing bottle will also have been noticed. The sulphur liberated combines with the ammonium sulphide to produce soluble polysulphides, $(NH_4)_2S_x$, where x may have almost any integral value between 1 and 9, the solution developing a deep yellow colour. After a time a deposit of sulphur may be seen on the bottom of the bottle.

The action of an acid on a polysulphide is to liberate free sulphur, with evolution of hydrogen sulphide, as can be shewn by acidifying a little yellow ammonium sulphide:

$$(NH_4)_2S_x + 2HCl = 2NH_4Cl + H_2S + (x - 1)S.$$

Oxides of Sulphur. When sulphur burns in air or oxygen the dioxide, SO_2, is formed, together with small quantities of the trioxide, SO_3. The former is not found naturally to any great extent. It occurs in volcanic gases, and traces of it are found in the atmosphere of towns. The odour is often noticeable near coke fires, as gas coke is rarely quite free from sulphur.

The sulphur dioxide required for the manufacture of sulphuric acid is usually made either by burning brimstone or by roasting iron pyrites, FeS_2,

$$4FeS_2 + 11O_2 = 2Fe_2O_3 + 8SO_2.$$

In the laboratory, it is usually obtained either by the action of a dilute acid on a sulphite:

$$Na_2SO_3 + 2HCl = 2NaCl + H_2O + SO_2,$$

or by the reduction of hot concentrated sulphuric acid by a metal, usually copper (Fig. 30). The equation is generally written:

$$Cu + 2H_2SO_4 = CuSO_4 + 2H_2O + SO_2,$$

but since black cuprous sulphide, Cu_2S, is produced at the same time, this equation is incomplete. Probably the first action of the metal on the acid is to displace hydrogen as in the case of the dilute acid with zinc, and this hydrogen reduces the sulphuric acid, mostly to sulphur dioxide, but in part to hydrogen sulphide. As has been seen, concentrated sulphuric acid is not difficult to reduce (cf. hydrogen iodide, p. 101) and even the passage of hydrogen through it, when heated, yields sulphur dioxide. The gas may be liquefied at atmospheric pressure by a mixture of ice and salt. Sulphur dioxide is usually sold in liquid form in thick glass syphons (see Fig. 31) of the size and shape of those used for soda water, and which contain from 4 to 5 lbs. of the liquid, the internal pressure being about two atmospheres at laboratory temperature. As can be seen from the molecular formula, the gas has about twice the density of air. It is poisonous even in dilute form if breathed for long, and is especially destructive to plants. At high temperatures, and also at ordinary temperatures under the influence of light, it dissociates into sulphur trioxide and free sulphur:

$$3SO_2 \rightleftharpoons 2SO_3 + S.$$

If a beam of light from an arc lamp be passed through a tube or cylinder of glass filled with sulphur dioxide, a mist, said to be due to free sulphur, forms in the gas. This disappears on standing in the dark.

When sulphur is burnt in oxygen, the volume of the sulphur dioxide produced is equal to the volume of the oxygen used; in other words, sulphur burning in oxygen causes no change in the gaseous volume, all the volumes

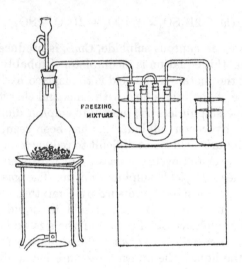

FREEZING MIXTURE

Fig. 30

being measured, of course, under the same conditions of temperature and pressure. Applying Avogadro's hypothesis, one molecule of oxygen produces one molecule of sulphur dioxide, the molecular formula being therefore S_xO_2. Since the density of the gas is 32 on the hydrogen standard, the molecular weight must be 64, and as the atomic weight of sulphur is 32, there can be but one atom of it present in the molecule of sulphur dioxide. (Cf. Hydrogen sulphide, p. 189.)

Sulphur dioxide is used technically as a bleaching agent for substances and fabrics which would be injured by chlorine, such as silk and wool. The articles to be bleached are hung up, usually in a slightly moist condition, in air-tight chambers in which brimstone is burnt. It has been suggested that the sulphur dioxide combines with the colouring matter to form a colourless compound, which in most cases is not very permanent. Woollen goods which have been bleached by sulphur dioxide regain their colour after a time, either by exposure to light and air, or by washing with soap, which has a slightly alkaline reaction. Articles bleached by this gas may often be restored to their original colour by treatment with alkalis. Sulphur dioxide has also the property of arresting the action of many ferments, and of destroying bacteria and vermin, and is on this account used for disinfecting rooms and clothing after cases of infectious disease. Wine and beer barrels are freed from injurious ferments by treatment with the gas before being refilled. Its employment for arresting the putrefaction of meat is now illegal.

The coefficient of solubility of sulphur dioxide in water at 0° C. is 80, and at laboratory temperature about 40. The solution has an acid reaction and conducts electricity, and is believed to contain sulphurous acid, to which, from a consideration of its salts, the formula H_2SO_3 is assigned. By cooling saturated solutions of the gas in water, several crystalline compounds, which are regarded as hydrates of the acid, $H_2SO_3 . nH_2O$, have been obtained. They all rapidly decompose into sulphur dioxide and water.

Two classes of salts derived from sulphurous acid are known; normal salts of the type M'_2SO_3, and acid salts or bisulphites, $M'HSO_3$. As a rule sulphites are rather insoluble in water, and when they do dissolve they are hydrolysed and give an alkaline reaction. On treatment

with acids they evolve sulphur dioxide, and decompose
when heated with formation of the metallic sulphide and
sulphate:

$$4K_2SO_3 = K_2S + 3K_2SO_4.$$

Sulphur dioxide does not in general support combustion.
Indeed a comparatively small proportion of it in air has
a marked effect in suppressing ordinary combustion and
extinguishing fires, yet it can oxidize in some cases. A
burning piece of magnesium plunged into the gas continues to burn, forming the oxide and the sulphide of the
metal. The oxidation of hydrogen sulphide by sulphur
dioxide has been mentioned. More usually it behaves as
a reducing agent, as also do sulphurous acid and sulphites.
Ferric salts in solution are reduced to the ferrous condition
on boiling with sulphurous acid:

$$Fe_2(SO_4)_3 + H_2SO_3 + H_2O = 2FeSO_4 + 2H_2SO_4.$$

A solution of mercuric chloride, when boiled with sulphurous
acid, gives a precipitate of mercurous chloride:

$$2HgCl_2 + H_2SO_3 + H_2O = 2HgCl + H_2SO_4 + 2HCl.$$

The halogens are converted into their hydracids:

$$Cl_2 + H_2O + H_2SO_3 = H_2SO_4 + 2HCl,$$

this being the action of sulphurous acid or sulphite when
used as an 'anti-chlor'; that is, for removing the odour
from fabrics which have been bleached with chlorine.
Iodine is taken into solution as hydriodic acid by sulphurous acid:

$$I_2 + H_2O + H_2SO_3 \rightleftharpoons H_2SO_4 + 2HI.$$

The conditions under which this reaction is reversible
have already been described.

When sulphur burns in air or oxygen, a little sulphur trioxide, SO_3, is said to be formed along with the dioxide, and the union of this trioxide with the moisture of the air is held to account for the fog observed when brimstone burns. If a mixture of sulphur dioxide with oxygen or air be passed over heated platinum, or other suitable catalyst, the dioxide is oxidized to the trioxide, which can be separated by passing the products of the reaction through a cooling apparatus, when long silky needle-like crystals are formed (Fig. 31). These melt at 14·8° C. to a clear liquid of boiling point 48° C.

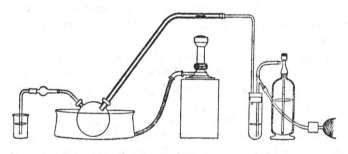

Fig. 31

Sulphur trioxide, or sulphuric anhydride, unites with water with great energy, and as the main problem in the manufacture of sulphuric acid is the production of cheap sulphur trioxide, this has received much attention. The conversion of sulphur dioxide into trioxide has been carried out technically by two processes, the later of which, the 'contact' process, is now almost exclusively used. The conditions which determine the course of the reversible reaction:

$$2SO_2 + O_2 \rightleftarrows 2SO_3$$

are temperature, pressure, and the relative proportions of

the sulphur dioxide and oxygen. As the formation of sulphur trioxide from the dioxide is an exothermic process, in order to obtain the maximum yield the temperature should be kept low, in accordance with the principle of Le Chatelier, but a low temperature means a slow reaction and therefore a small yield. The optimum temperature, that is, the most efficient from the technical point of view, is found by experiment to be about 400° C. when an effective catalyst is used.

As the volume of the sulphur trioxide is two-thirds of the volume of the reacting gases, increase of pressure should increase the yield of trioxide. In actual practice, however, the process works efficiently at ordinary atmospheric pressures.

In the earlier experiments on the direct union of sulphur dioxide and oxygen, it was observed that the best results were obtained when the reacting gases contained excess of oxygen beyond that required by the chemical equation. The necessity for using excess of oxygen can be shewn theoretically. The Law of Mass Action, first clearly enunciated by Guldberg and Waage in 1867, states that *the rate of chemical change is proportional to the active mass of each of the reacting substances.* By active mass is to be understood molecular concentration; that is, the number of molecules in a given volume. Applying this law, the rate of formation of sulphur trioxide is proportional to the *square* of the concentration of the sulphur dioxide, and to the *first power* of the concentration of the oxygen, while the rate of the reverse reaction is proportional to the *square* of the concentration of the sulphur trioxide. At equilibrium these rates are equal. The efficiency of the process; that is, the fraction of sulphur dioxide converted to sulphur trioxide, is therefore proportional to the *square root* of the oxygen concentration. The following numbers were given

by a mixture of sulphur dioxide and air reacting at a
temperature of 400° C. in presence of a catalyst:

Composition of reacting gases in percentages by volume.			Percentage of SO_2 converted into SO_3.
N_2	SO_2	O_2	
83·0	7·0	10·0	99·3
80·0	2·0	18·0	99·5

For a long time the contact process was not a commercial success, largely owing to the short life of the expensive catalyst. The sulphur dioxide contained *something*
which gradually diminished the efficacy of the catalyst
below the point of economic working. The enormous advance which has taken place in recent years is mainly due
to improvement in the methods of removing the impurities
which 'poison' the catalyst. Arsenic compounds are particularly injurious, and since arsenic is found in nearly all
metallic sulphide ores, the sulphur dioxide made from iron
pyrites requires very careful purification. This is effected
by passing the gases from the pyrites burners through
chambers into which steam is injected. The fine particles
of dust and arsenious oxide provide nuclei on which the
steam condenses, and they are thus washed out. A similar
operation is subsequently carried on at a lower temperature,
the gas being passed up towers down which fine sprays of
water fall. Finally it is thoroughly dried by passage
through a sulphuric acid scrubber before entering the contact chamber; a cylindrical apparatus containing a number
of tubes in which platinized asbestos is placed between
perforated shelves. This chamber is heated externally at
the commencement of operations, but since much heat is
evolved by the reaction itself, the flow of gas has to be
carefully regulated so as to maintain the temperature at
about 400° C. Pumice coated with precipitated ferric oxide

is sometimes used instead of platinized asbestos, but is not so efficient. Although sulphur trioxide unites with water energetically, the absorption presents difficulties, since the trioxide assumes the spheroidal condition and the tiny droplets are then slow in dissolving. The best absorbent is found to be ordinary concentrated sulphuric acid of about 98 per cent. strength. The gases are passed into tanks containing this acid, and at the same time an equivalent quantity of water is admitted. The 'contact' works produce acids in which the proportion of sulphur trioxide to water has almost any value from 98 per cent. H_2SO_4 to nearly pure trioxide. The technical product known as 'oleum' is not a chemical individual, but a mixture which may be regarded as sulphuric acid + sulphur trioxide.

Compounds of Sulphur Trioxide and Water. When sulphur trioxide and water (or sulphuric acid and water) are mixed, the heat-concentration curve is continuous; that is, there are no breaks indicating the formation of definite compounds. On cooling solutions of varying concentration, maximal freezing points have been noted, corresponding to definite hydrates, such as $H_2SO_4 . H_2O$ and $H_2SO_4 . 4H_2O$, but these are not important here. If ordinary concentrated sulphuric acid be cooled in a freezing mixture of ice and salt, crystals separate which can be drained off from the more dilute acid which remains liquid, and examined. They melt at 10° C., and correspond to the monohydrate of sulphur trioxide, $SO_3 . H_2O$, or H_2SO_4. Pure sulphuric acid can therefore be obtained in this way. Another hydrate, $2SO_3 . H_2O$, or $H_2S_2O_7$, can be separated by the addition of sulphur trioxide to ordinary sulphuric acid in approximately the calculated quantity, freezing and draining as before. These crystals melt at 35° C., the compound being known as pyrosulphuric acid. Formerly a fuming kind of sulphuric acid was made at Nordhausen, in

Saxony, by distilling green vitriol (crystallized ferrous sulphate). This was known as Nordhausen acid, and contained about 16 parts of sulphur trioxide to 84 parts of sulphuric acid, being probably a mixture of sulphuric and pyrosulphuric acids.

For many years sulphuric acid was manufactured exclusively by what is known as the lead chamber process, and although this is now obsolescent, owing to the success of the contact process, the chemistry of the older method is still important. As in the case of the newer process, the main problem is the efficient oxidation of sulphur dioxide to trioxide, and this is brought about by the catalytic action of oxides of nitrogen. The stages of the operation are usually represented, at least qualitatively, thus:

The hot mixture of sulphur dioxide and air from the pyrites burners passes over crucibles in which nitric acid is being liberated, sweeping the fumes of the acid into a large chamber made of sheet lead. The nitric acid fumes oxidize some of the sulphur dioxide:

$$4HNO_3 = 2H_2O + 4NO + 3O_2.$$

$$3O_2 + 6SO_2 = 6SO_3.$$

The nitric oxide is oxidized by the air to nitric peroxide·

$$2NO + O_2 = 2NO_2.$$

The nitric peroxide oxidizes sulphur dioxide:

$$SO_2 + NO_2 = SO_3 + NO,$$

and so the cyclic action continues.

The sulphur trioxide is converted into sulphuric acid by blowing jets of steam into the chamber at various points. The liquid which settles is known as chamber acid, and usually has a concentration of from 60 to 70 per cent. of H_2SO_4.

As in all catalytic actions, the *quantity* of oxidation possible is not a function of the mass of oxides of nitrogen

present, but the *velocity* of the reaction is, and on this
account relatively large amounts of the somewhat costly
nitric acid are used. As much unwanted atmospheric
nitrogen has to pass through the chamber, the issuing
waste gases carry considerable quantities of oxides of
nitrogen with them. These are efficiently recovered by a
scrubber invented by Gay-Lussac in 1827, which consists
of a tall tower packed with coke or flints over which concentrated sulphuric acid trickles. The spent gases are
passed up this, the oxides of nitrogen being absorbed by
the sulphuric acid and forming what is known as 'nitrated'
acid or nitrous vitriol. This may be regarded as sulphuric
acid in which one of the hydrogen atoms of the molecule
has been replaced by a nitrosyl group (NO), the compound
being hydrogen nitrosyl sulphate, or nitrosyl sulphuric
acid.

$$O_2S\Big\langle{{OH}\atop{OH}} \longrightarrow O_2S\Big\langle{{OH}\atop{O\,(NO)}}$$

If the oxides of nitrogen obtained by warming nitric acid
with arsenious oxide be passed through concentrated
sulphuric acid, this compound separates in crystalline form.
It is sometimes seen forming in the lead chamber itself
(through observation windows) and is known technically
as chamber crystals. These crystals have been thought to
play an important part in the production of sulphuric acid
within the chamber, but the point is doubtful. The reaction,
which is reversible, may be represented:

$$2H_2SO_4 + NO + NO_2 \rightleftharpoons 2H(NO)SO_4 + H_2O,$$

excess of water reacting with the crystals to reproduce
sulphuric acid and oxides of nitrogen.

The liquid from the bottom of the Gay-Lussac tower is
forced to the top of another and somewhat similar struc-

ture, the Glover tower, placed between the pyrites burners and the lead chamber. The 'nitrated' acid passes down this, mixing on its way with chamber acid, the water from which reacts with it to reform sulphuric acid and set free oxides of nitrogen. The gases from the burners passing up the tower sweep these back into the chamber, while the acid accumulating at the foot of the Glover tower is more concentrated. No doubt sulphuric acid is formed inside the Glover tower itself, as the sulphur dioxide and air meet the

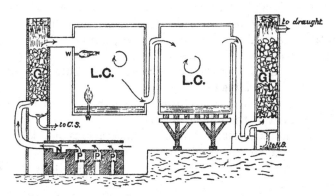

Fig. 32

liberated oxides of nitrogen. Where the Glover tower is used, it is customary to send *all* the chamber acid produced through it, the acid collected at the foot usually containing about 80 per cent. of H_2SO_4. This is sometimes sold without further treatment as 'brown oil of vitriol,' known technically as B.O.V.

One of the most troublesome and expensive operations in this manufacture is the concentration of the acid by evaporation. Formerly this was done, firstly in leaden vessels which are only attacked by somewhat concentrated acid, and finally in stills made of platinum. Latterly glass vessels have replaced the platinum stills. When *dilute* sulphuric acid is subjected to distillation, the volatile product is almost pure water. As this is removed the concentration increases, until at a temperature of about 338°C. the distillate has the same composition as the residue, and further concentration cannot be obtained in this way. On the other hand, when *pure* sulphuric acid is distilled sulphur trioxide is evolved, the water being retained by the rest of the acid until the same concentration is reached. This constant boiling mixture contains about 98 per cent. of acid and 2 per cent. of water (cf. hydrochloric and nitric acids). As has been mentioned, pure sulphuric acid may be obtained from this by freezing; a remarkable phenomenon in view of the heat evolved when sulphuric acid and water are mixed.

The percentage composition of sulphuric acid can be determined in the following way:

1. A known weight of pure sulphuric acid is diluted with water, and an excess of zinc or magnesium added. The hydrogen evolved is measured. This gives the percentage of hydrogen.

2. A known weight of galena (lead sulphide, PbS) is oxidized by evaporating to dryness several times with nitric acid until the weight is constant. The sulphide is converted into sulphate. This gives the ratio $PbS : PbSO_4$.

3. A known weight of sulphuric acid is diluted with water, and excess of a solution of lead nitrate added. Lead sulphate is precipitated and can be filtered off, washed, dried, and weighed. This gives the ratio $H_2SO_4 : PbSO_4$.

4. A known weight of metallic lead is covered with concentrated sulphuric acid and evaporated to dryness, this operation being repeated until the weight of the residue is constant. This gives the ratio $Pb : PbSO_4$.

The vapour density of sulphuric acid at 450° C. is only

about one-half of that required by the formula H_2SO_4. This points to practically complete dissociation of the acid into sulphur trioxide and water at 450° C.

Sulphuric acid can form two (and only two) kinds of salts with monovalent metals like sodium and potassium. This indicates the presence of two (and only two) atoms of hydrogen in the molecule. Hence the molecular formula of sulphuric acid is written H_2SO_4.

Sulphuric acid is a colourless, oily liquid of specific gravity 1·84. It is therefore nearly twice as heavy as water, and when this fact is appreciated it should help the student to avoid one of the commonest and most dangerous mistakes made in chemical laboratories; that of using the *concentrated* acid in cases where the *dilute* acid is required. When the concentrated acid is mixed with water much heat is evolved, and a contraction in volume takes place. The mixing should be done by pouring the acid into water, and never water into the acid. Some of the uses of the concentrated acid as a desiccating agent have already been described. A beaker containing 100 grm. of the ordinary acid, left exposed in the laboratory for a week, absorbed 50 grm. of water from the air. The elements of water can sometimes be removed from substances which do not actually contain water *as such*, as has been described in the cases of alcohol and oxalic and formic acids. Carbohydrates, that is, compounds containing carbon, together with hydrogen and oxygen in the proportions in which they are found in water, and which can be represented generally by the formula $C_xH_{2y}O_y$, are 'dehydrated' and charred by the concentrated acid, as can be seen by pouring a little of it upon cane sugar, starch, wood, paper, straw, cotton, or similar material. Hydrocarbons of the paraffin series are either wholly unattacked, or only very slightly affected, even by the most concentrated acid. Almost all the commoner metals react with concentrated sulphuric acid in the manner described in the case of copper. It will have been noted that whenever *dilute* sulphuric acid reacts with a metal, hydrogen is the only gas evolved, and that with *concentrated* acid the evolved gas is sulphur dioxide; never hydrogen. Further, that twice as much sulphuric acid takes part in the reaction when it is concentrated as when dilute, as a comparison of the equations which follow will shew.

$$Zn + H_2SO_4 = ZnSO_4 + H_2 \quad \text{(dilute)},$$

$$Zn + 2H_2SO_4 = ZnSO_4 + 2H_2O + SO_2 \quad \text{(concentrated)}.$$

Sulphuric acid, like other acids, reacts with basic oxides to produce a salt and water only. When warmed with metallic peroxides, it yields a sulphate corresponding to a lower oxide of the metal, water, and oxygen (cf. hydrochloric acid).

$$2MnO_2 + 2H_2SO_4 = 2MnSO_4 + 2H_2O + O_2.$$

Classification of Salts. Salts are usually classified as normal salts, acid salts, basic salts, and double salts*.

A *normal* salt may be defined as the compound formed when the whole of the acid hydrogen of an acid is replaced by a metal (or a radical acting as a metal), as in potassium sulphate, K_2SO_4, and ammonium sulphate, $(NH_4)_2SO_4$.

An *acid* salt is the compound obtained by replacing the acid hydrogen of an acid only *partly* by a metal, as in potassium hydrogen sulphate (potassium bisulphate or acid potassium sulphate), $KHSO_4$. It should be noted that the term *acid* refers to the composition of the salt, and not to its reaction with litmus or other indicator. Potassium bicarbonate, $KHCO_3$, for instance, is an *acid*

* 'Complex' salts are discussed in Chapter x. See also p. 208.

salt with an *alkaline* reaction. It will be obvious that monobasic acids; that is, acids which contain but one atom of acid hydrogen in the molecule, like nitric acid, cannot form acid salts, though they can and do form *normal* salts with an *acid reaction*.

A *basic* salt is the compound formed by the union of a normal salt with its own base. This will be better understood after trying the following test tube experiments.

If to a little cold solution of copper sulphate, cold caustic soda solution be added, a blue precipitate of copper hydroxide will be formed:

$$CuSO_4 + 2NaOH = Cu(OH)_2 + Na_2SO_4.$$

If this be boiled, the soda being in excess, the hydroxide will lose water and yield the black oxide, and this does not hydrate again on cooling:

$$Cu(OH)_2 \longrightarrow CuO + H_2O.$$

If a few drops of caustic soda solution be added to half a test tube of copper sulphate solution (that is, the copper sulphate being in large excess) the blue precipitate at first formed does *not* turn black on boiling, or only does so in spots, which soon disappear as the boiling continues. A heavy greenish precipitate settles out from the still blue solution. This is a basic sulphate of copper; a compound of copper sulphate with copper oxide, $(CuSO_4)_x.(CuO)_y$, where x and y are small integral numbers. A similar insoluble basic copper sulphate is sometimes formed when a hurried attempt is made to dissolve copper sulphate crystals in hot water. When solutions of sodium carbonate and copper sulphate are mixed, the precipitate is not copper carbonate, as might be expected, but the *basic* carbonate, $(CuCO_3)_x.(CuO)_y$. Metal work in which copper predominates, in time acquires a patina of basic copper carbonate when exposed to the air. Verdigris is the basic *acetate* of copper*.

The classification may be briefly summarized in this way:

In a normal salt, the number of equivalents of the acid = number of equivalents of base.

In an acid salt, the number of equivalents of the acid > number of equivalents of base.

In a basic salt, the number of equivalents of the acid < number of equivalents of base.

Basic salts are in general but little soluble in water, though the basic acetate of lead, which will be described later, is an important exception.

If a few grammes of potassium chlorate be heated to low redness for a short time, cooled, dissolved in water, and the solution allowed to evaporate to crystallization, examination with a lens will shew the presence of more than one kind of crystal. By careful picking, the chloride, the chlorate, and the perchlorate of potassium can be separated from the mixed mass of crystals, each being quite separate and distinct.

If solutions of magnesium sulphate and ferrous sulphate be mixed and allowed to crystallize, a different result is obtained. On picking out a single crystal, it will be found to contain both magnesium and iron, probably in a proportion which bears no relationship to the respective atomic weights of the metals themselves. Further, by taking differing proportions of the two salts originally, the composition of the crystals which separate can be varied. These are true 'mixed' crystals, and are only formed when their component salts are isomorphous.

---

* Many so-called basic salts, to which formulae have been assigned, are not chemical individuals at all, but *mixtures* of the normal salt with the oxide or hydroxide of the metal.

If about 20 grm. of crystallized ferrous sulphate and 10 grm. of ammonium sulphate be dissolved separately in water containing a little sulphuric acid, the two solutions mixed, and then allowed to evaporate until a few grammes of crystals are deposited, these on examination will be found to be neither ferrous sulphate nor ammonium sulphate, but something differing in appearance from both. They contain iron and ammonium in exactly equivalent proportions, though this is by no means the proportion in which iron and ammonium are present in the mother liquor. A true chemical compound of ferrous sulphate and ammonium sulphate has been formed; that is, a double salt.

A *double salt* may be regarded as the product obtained when two simple salts combine in definite proportions, as distinct from 'mixed' salts, where the proportions are indefinite and variable.

The 'vitriols' are the normal sulphates of certain bivalent metals which crystallize out from solution with seven molecules of crystal water, with the important exception of copper, which has only five. Crystallized copper sulphate, $CuSO_4 . 5H_2O$, is known as blue vitriol, or bluestone. Zinc sulphate, $ZnSO_4 . 7H_2O$, is white vitriol, and ferrous sulphate, $FeSO . _4 7H_2O$, is green vitriol, or copperas. These vitriols seem to be able to exchange one of their molecules of crystal water for the sulphate of an alkali metal, forming with it a 'double vitriol' such as ferrous ammonium sulphate, $FeSO_4 . (NH_4)_2SO_4 . 6H_2O$. This compound is especially important in analytical work, since it contains ferrous iron in a form in which it is not liable to oxidation by the air; and since the crystals are definite and permanent, standard solutions can be made from them by direct weighing. As observation will shew, they contain one-seventh of their weight of iron. Potassium ferrous sulphate is a similar salt, $K_2SO_4.FeSO_4.6H_2O$. The double salts formed by copper are quite regular, in so far as they all have six molecules of crystal water like the rest.

$$(CuSO_4.K_2SO_4.6H_2O.)$$

These double vitriols are isomorphous, and single 'mixed' crystals containing a considerable number of metals, in indefinite proportions, may be obtained by mixing their solutions and allowing them to crystallize.

Another series of double salts, the alums, may be regarded as double sulphates of the type:

$$M'_2SO_4.M''''_2(SO_4)_3.24H_2O,$$

where M' is a univalent metal, and M'''' tervalent. When M' and M'''' are not specially named, potassium and aluminium respectively are understood. Thus 'alum' is $K_2SO_4.Al_2(SO_4)_3.24H_2O$; ammonium iron alum is

$$(NH_4)_2SO_4.Fe_2(SO_4)_3.24H_2O;$$

and chrome alum $K_2SO_4.Cr_2(SO_4)_3.24H_2O$.

Since nothing is really known of the molecular weights of these compounds they can equally well be written

$$M'.M''''(SO_4)_2.12H_2O.$$

The alums also form an isomorphous family, and easily make both 'mixed' crystals and 'layer' crystals; that is, a crystal of chrome alum will 'grow' in a saturated solution of alum.

Potassium ferrocyanide, $K_4FeC_6N_6$, is sometimes erroneously written as though it were a double salt of potassium cyanide and ferrous cyanide, $(KCN)_4.Fe(CN)_2$, but it does not shew the ordinary reactions of iron as a double salt of iron and potassium should. For instance, no precipitate is given with ammonia or soda. It is more correctly regarded as the potassium salt of ferrocyanic acid, $H_4FeC_6N_6$, where the iron is part of the acid radical (FeC_6N_6). Salts which contain more than one metal and yet produce but one metallic cation in solution, like ferrocyanides and ferricyanides, are usually classified as 'complex.' (See Chap. x.)

Thiosulphates. When aqueous sodium sulphite is boiled with powdered sulphur, the latter is taken into solution. On filtering off the excess of sulphur and evaporating, sodium thiosulphate* separates in prismatic crystals which contain five molecules of crystal water, $Na_2S_2O_3.5H_2O$. Since solutions of this compound readily dissolve the halides of silver, it is used in photography, under the name hyposulphite, or hypo, for fixing pictures; which means the removal of the silver halide left unchanged in the plate or print after exposure and development:

$$2AgCl + 3Na_2S_2O_3 = Ag_2Na_4(S_2O_3)_3 + 2NaCl,$$

or, with formation of a rather less soluble salt:

$$AgCl + Na_2S_2O_3 = AgNaS_2O_3 + NaCl.$$

It is also used to some extent for extracting silver from silver ores when the metal is present as the chloride.

Iodine is converted into sodium iodide by thiosulphate, the latter being changed into sodium tetrathionate:

$$2Na_2S_2O_3 + I_2 = 2NaI + Na_2S_4O_6.$$

This property is used in volumetric analysis for determining the weight of iodine present in, or liberated by, a reaction. A similar though not identical reaction takes place between chlorine and sodium thiosulphate, and on this account the latter sometimes replaces sodium sulphite as an antichlor in bleaching works. Thiosulphuric acid is unstable and has

---

* The prefix 'thio-' signifies the substitution of an atom of sulphur for an atom of oxygen in the molecule. Thus sulphate = $M''SO_4$; thiosulphate = $M''S_2O_3$. Cyanate = $M'CNO$; thio-cyanate = $M'CNS$. Urea = $(NH_2)_2CO$; thio-urea = $(NH_2)_2CS$.

not been isolated. When a solution of sodium thiosulphate is acidified with dilute hydrochloric or sulphuric acid, it remains clear for a few seconds and then develops a turbidity, which rapidly resolves into a precipitate of sulphur, sulphur dioxide being at the same time liberated:

$$H_2S_2O_3 = H_2O + S + SO_2.$$

Sodium thiosulphate in which silver chloride has been dissolved, yields on acidifying a black precipitate of silver sulphide, Ag_2S.

Persulphuric Acid. A mixture of oxygen and sulphur dioxide, submitted to the silent electric discharge in an ozonizing apparatus, yields oily drops of an unstable oxide of sulphur, S_2O_7, the anhydride of persulphuric acid. The acid itself is formed when somewhat concentrated sulphuric acid is electrolysed, only traces of oxygen being liberated. It is also produced by the action of hydrogen peroxide on concentrated sulphuric acid, which is one of the general methods for the preparation of per-acids. The potassium salt is easily prepared by the electrolysis of a cold saturated solution of potassium bisulphate, the sparingly soluble persulphate being deposited at the anode. The crystals can be filtered off, washed with cold water and dried. They are fairly stable, but on heating they break down into oxygen and potassium pyrosulphate:

$$2K_2S_2O_8 = 2K_2S_2O_7 + O_2.$$

At a higher temperature the pyrosulphate decomposes into potassium sulphate and sulphur trioxide. Persulphates are nearly all soluble, and are good oxidizing agents. They liberate iodine from solutions of iodides, and precipitate hydrated manganese dioxide from solutions of manganous salts on heating.

Carbon Disulphide. Carbon at a red heat unites directly

with the vapour of sulphur to form carbon disulphide, CS_2, the reaction being endothermic:

$$C + S_2 = CS_2 - 22 \cdot 3 \ K. \quad \text{(for the liquid)}.$$

This compound is manufactured on quite a large scale in electrothermal furnaces, which are cylindrical towers about 40 feet high with a diameter of 15 or 16 feet, and are filled with coke. The arc passes between carbon electrodes placed near the base of the tower, and gradually raises the whole mass to the required temperature. Melted sulphur is fed in at the foot of the tower where it is vaporized, and the vapour made to pass over a large surface of red hot coke. The carbon disulphide produced is led into long cooling coils where it is condensed. It is then purified by several redistillations, and by violent shaking with mercury to remove dissolved sulphur.

Carbon disulphide is a rather heavy, colourless, highly refractive, and very inflammable liquid, boiling at 46° C., and only solidifying at very low temperatures. The commercial product has an unpleasant smell, which disappears on purification. While mixing in all proportions with alcohol, ether, and benzene, it is almost insoluble in water. It readily dissolves iodine, phosphorus, sulphur, india-rubber, resins, waxes, fats, and oils. It is used for the final extraction of oil from crushed seeds and fruits, and for extracting perfumes from flowers.

CHAPTER VII

PHOSPHORUS AND ITS COMPOUNDS

PHOSPHORUS is not found in nature in the uncombined
state, but in combination with calcium and oxygen it occurs
abundantly in bones and guano, and in minerals such as
phosphate rock, apatite, coprolites, etc. Phosphates are
necessary for the growth of both plants and animals, and
are found in all fertile soils. The liquid products of animal
metabolism contain phosphates, mainly in the form of the
sodium or ammonium salts, and these are in great measure
returned to the soil. Fresh bones consist of more than half
their weight of calcium phosphate, and about 15 lbs. of this
compound is contained in the average human skeleton.

Bones may be regarded as consisting of calcium phosphate and cartilaginous matter, and are variously treated
according to the materials it is required to obtain from
them. By extraction with organic solvents such as benzene,
carbon tetrachloride, or acetylene tetrachloride, the fat can
be removed and recovered. The bones, so cleaned, yield
gelatine when heated with water under pressure. After
this treatment they have lost their smooth glossy appearance, are chalk like, and adhere to the tongue, little being
left but calcium phosphate and a small percentage of
carbonate. Heated out of contact with the air, the cartilaginous matter of bones undergoes destructive distillation, yielding water, ammonia, and a tarry liquid with
a burning taste known as Dippel's oil, from which pyridine
is obtained. The bones retain their shape under this treatment, but turn black from the presence of carbon remaining
from the destruction of the organic matter. This can be
freed from the calcium phosphate by boiling with hydro-

chloric acid, when bone black or bone charcoal remains. When bones, either before or after the degelatinizing process, are heated strongly in the presence of air, all organic matter burns away and bone ash remains, and if all traces of flesh and blood are removed before incineration, this is quite colourless.

Phosphorus is obtained from calcium phosphate either by the retort process or by the electro-thermal method. In the former, now almost obsolete, calcium phosphate is boiled with the calculated quantity of sulphuric acid required to convert all the calcium into the sparingly soluble sulphate:

$$Ca_3(PO_4)_2 + 3H_2SO_4 = 3CaSO_4 + 2H_3PO_4.$$

The soluble phosphoric acid is then strained away, mixed with powdered carbon (coke or charcoal) and evaporated to a cindery mass in iron pans. This is then packed into iron or clay retorts with long necks, each holding about 7 lbs. These are arranged in tiers in a coke furnace, the necks projecting through holes in the furnace wall, and being connected to down-cast pipes dipping under water. On heating, the phosphoric acid loses water and becomes meta-phosphoric acid,

$$H_3PO_4 = HPO_3 + H_2O.$$

At a higher temperature, the metaphosphoric acid is reduced by the carbon:

$$4HPO_3 + 12C = 2H_2 + 12CO + P_4,$$

the volatilized phosphorus being condensed under the water.

In the electric furnace method the calcium phosphate (in this case usually obtained from mineral sources and not from bones), is mixed with powdered coke and silica, SiO_2, in the form of fine sand, and heated to a high temperature by the electric arc. The reaction may be regarded as an exchange of the silica for the phosphoric anhydride, and a reduction of the latter by the carbon:

$$Ca_3(PO_4)_2 + 3SiO_2 = 3CaSiO_3 + P_2O_5,$$

$$2P_2O_5 + 10C = P_4 + 10CO,$$

the phosphorus vapour being led away and condensed as before.

As obtained by either of these methods, phosphorus is chocolate coloured and impure. The technical methods of purification are regarded as trade secrets, but they probably consist in an oxidation process of some kind, followed possibly by a second distillation. When impure phosphorus is boiled with potassium bichromate solution, in presence of a little dilute sulphuric acid, there is but little loss of phosphorus, and the product is an almost colourless liquid. In this condition it may be run into glass tubes in which it solidifies, at the same time contracting slightly, and so is easily detachable from the tube.

Allotropy of Phosphorus. Various modifications of phosphorus have been described, the existence of some of which at least cannot be said to be beyond doubt. The two best-known forms are the ordinary kind, described by many names, such as white, common, vitreous, etc., and the dark chocolate coloured variety, known as safety, red, or 'amorphous' phosphorus. White phosphorus is metastable under all conditions, and passes into the red modification from the instant it is made, at a rate which, like that of most chemical changes, approximately doubles for each rise in temperature of 10° C. Even at ordinary laboratory temperatures sticks of white phosphorus soon lose their translucent appearance, and red patches develop on the surface, especially when exposed to light. The change can be accelerated by certain catalysts; by a trace of iodine, for instance. Since the change:

P (white) → P (red) + 3·7 K. per gramme-atom

proceeds in one direction only, the element is monotropic; that is, there is no transition point, such as has been described in the case of the two crystalline forms of sulphur. The only way in which red phosphorus can be converted into white is by vaporizing it and quickly condensing the vapour. Whenever a solid is condensed from a vapour, or crystallized from solution, the least stable modification is the first to appear.

The less stable modification of a substance has always the greater vapour pressure and the greater solubility. In the case of phosphorus, the vapour pressure of the white variety increases with rise of temperature much more rapidly than that of the red; so much more so, that the vapour pressure of the former at 200° C. is greater than that of the latter at 350° C. It should therefore be possible to distil white phosphorus from the *colder* part of a tube and condense the vapour as red phosphorus at the *hotter* part. Such a distillation was actually carried out by Troost and Hautefeuille, between the temperatures 324° C. and 350° C.

The manufacture of red phosphorus from white is carried out in a vessel of iron or earthenware, provided with a wide mouth which can act as a safety valve, and which is closed with an air-tight cover as soon as most of the air has been expelled. The vessel with its charge of white phosphorus (often a hundredweight or more) to which a trace of iodine has been added, is maintained at a temperature of 230°– 235° C. for some time. If the temperature be allowed to rise much above 235° C. the change from white phosphorus to red takes place too energetically for safety. When the operation has proceeded far enough, the hard reddish mass remaining is ground up under water, and the unchanged

white phosphorus removed. Since the latter is soluble in
carbon disulphide, in which red phosphorus does not dissolve, extraction with this solvent leaves red phosphorus
very free from white, but on a large scale it is more economical to boil the product with aqueous soda, which while
readily attacking white phosphorus, is almost without
action on red, unless concentrated. The reaction is described below.

White phosphorus has a specific gravity of 1·83. At
0° C. it is hard, brittle, and crystalline; at ordinary temperatures it is softer, and somewhat wax-like. It melts at
44° C., and boils at 290° C., but can be made to sublime
without melting at about 40° C. The ignition temperature
is very low (50° C. in moist air) and therefore white phosphorus is usually kept under water. The vapour is poisonous,
and on this account it is no longer used in the making of
matches in this country. Taken internally, the solid is
very toxic. Poison pastes containing it are sold for the
destruction of beetles, rats, and other vermin. White phosphorus is luminous in the dark; even minute traces of it,
when boiled in water, imparting a luminosity to the issuing
steam. This phenomenon can be used as a sensitive de-
tector for the presence of white phosphorus in poisoning
and other cases. The substance to be tested is placed in a
flask half full of water. On boiling in the dark, a 'ghost
flame' is seen to play over the neck of the flask. This disappears if a little alcohol or turpentine be added to the
water.

Although practically insoluble in water, white phosphorus dissolves in ether, chloroform, benzene, turpentine, carbon disulphide, olive oil, and many other organic
solvents. If a solution in carbon disulphide be allowed to
evaporate slowly in an inert atmosphere, such as carbon
dioxide, octahedral crystals of white phosphorus can be

obtained. If this solution be poured on filter paper and exposed to the air, upon evaporation of the carbon disulphide the phosphorus is left on the paper in a fine state of division, and in this condition offers so much surface to the oxidizing action of the air that it inflames in a few seconds.

Although phosphorus is luminous in air, it shews no luminosity in pure oxygen, except at low pressures. Notwithstanding that the whole of the oxygen can be removed from an enclosed volume of air at ordinary temperature by the introduction of a stick of phosphorus, as is often done in gas analysis, yet no appreciable absorption of the gas takes place when phosphorus comes in contact with even approximately pure oxygen at atmospheric pressure. It is not practicable, for instance, to estimate the percentage of oxygen in the commercial gas in this way, unless a measured volume of an inert gas be introduced, or the pressure of the oxygen be reduced in some other way. It has already been mentioned that when phosphorus glows in moist air, sufficient ozone is produced to be identified.

Phosphorus combines directly with halogens, and with many metals to form phosphides. The presence of hardly more than a trace of phosphorus in a metal may modify its properties enormously. The introduction of a very small proportion of copper phosphide into bronze (making phosphor bronze), for instance, greatly increases its power of resisting corrosion by sea-water, and, on the other hand, greatly diminishes its conductance for electricity.

The reducing action of white phosphorus on certain metallic salts is remarkable. If a clean stick of it be placed in a solution of copper sulphate, in a short time the phosphorus is coated with a lustrous plating of metallic copper, and on standing (in excess of copper sulphate) all the phosphorus disappears, while an equivalent weight of copper in a granular condition is left at the bottom of the vessel. Similar reactions take place with solutions of silver and gold salts.

On boiling with nitric acid, even when dilute, phosphorus is converted into phosphoric acid.

Since the conversion of white phosphorus into red is an exothermal change, the red variety is the less energetic form. It is not amorphous. On being heated under pressure and allowed to cool, rhombic crystals are formed. It is denser than the white variety, its specific gravity being 2·14. Red phosphorus generally shews an acid reaction to indicators. This is due to the presence of traces of unconverted white phosphorus, which is slowly oxidized by the air, becoming ultimately phosphoric acid. Even when this is removed by exhaustive washing with water, the acid reaction returns after a time.

Red phosphorus is not poisonous when pure. It seems to pass through the alimentary system without undergoing any change. It shews no luminosity in the dark; does not take fire at temperatures below 260° C.; has no measurable vapour pressure at ordinary temperature, and is insoluble in all the ordinary solvents. It dissolves in phosphorus tribromide, a liquid boiling at 175° C., and in molten lead, from which it separates in crystals on cooling.

Red phosphorus is used in the manufacture of safety matches, or rather, of the surface on which they strike. Safety match heads usually carry a mixture of potassium chlorate and antimony sulphide, held together on the wood by gelatine, but no phosphorus. The striking surface is made by painting a mixture of red phosphorus and fine glass powder on the side of the box with a solution of gelatine.

Atomicity of the Phosphorus Molecule. Phosphorus combines with chlorine to form a liquid compound which can be purified without difficulty (see below). By interaction with silver nitrate in presence of water, this compound can be shewn to contain 77·42 per cent. of chlorine, and therefore 22·58 per cent. of phosphorus. The equivalent weight of phosphorus, i.e. the weight of it united with 35·45 grm. of chlorine, is therefore

$$\frac{22\cdot58 \times 35\cdot45}{77\cdot42} = 10\cdot34.$$

The atomic weight of phosphorus must be $10\cdot34 \times n$, where n is a small whole number.

As the density of the chloride when vaporized is found by experiment to be 68·6 times that of hydrogen, its molecular weight is 137·2; and 22·58 per cent. of this $= 31\cdot02$.

Therefore the atomic weight of phosphorus cannot be *greater* than 31·02. Since no compound of phosphorus has hitherto been found to contain *less* than 31 grm. of phosphorus in one gramme-molecule, the value of n is accepted as 3, and the atomic weight of phosphorus as 31 (very nearly).

The density of the vapour of phosphorus, from approximately 500° C. to 1000° C., is very nearly constant at about 62 times that of hydrogen; that is, the molecular weight between these temperatures is 124, indicating a tetratomic molecule, P_4. Above 1000° C. P_4 dissociates into diatomic molecules.

The molecular complexity of phosphorus in solution varies to some extent with the solvent, but here too, tetratomic molecules are indicated, at all events when carbon disulphide is the solvent.

Oxides of Phosphorus. Phosphorus forms several compounds with oxygen, of which only two need be considered here; the pentoxide, P_2O_5, and the trioxide, P_2O_3. These names and formulae were empirical in the first instance, but they have been retained for convenience, although the vapour densities of the compounds, over a wide range of temperature, are now known to correspond to the double formulae, P_4O_{10} and P_4O_6 respectively. When phosphorus is burned in air the pentoxide, which is the main product, appears in dense white clouds, and use has been made of this property in the Navy for producing masking screens of artificial fog at sea, by burning phosphorus. Small quantities of the pentoxide can be made in this way by conducting the burning under a bell jar standing on an earthenware plate. On a larger scale, phosphorus is burned in oxygen in an apparatus in which the process can be carried on semi-continuously. The only practicable way of obtaining phosphorus pentoxide is by direct union of the elements.

Even when the oxygen is in excess some of the lower oxide is always formed, and the proportion of this can be increased by checking the admission of oxygen. In the preparation of the trioxide, phosphorus is burned in a glass tube through which a slow stream of air passes. The pentoxide is arrested by a plug of glass wool placed further along the tube, where it is heated by a hot-water jacket, while the trioxide can pass on into a U-tube surrounded by cold water, where it condenses. When pure, phosphorus trioxide is a solid which melts at 22·5° C., and boils at 173° C.

The common impurity in phosphorus pentoxide is combined water (that is to say, phosphoric acid) and from this it can be purified by sublimation. When the trioxide is the impurity to be removed, the sublimation must be carried on in an atmosphere of oxygen and the vapours passed over a catalyst, such as platinum, before being condensed. The presence of phosphorus trioxide in the product obtained by burning phosphorus in air under a bell jar can be shewn by its reducing action. The white solid, when dissolved in water and boiled with silver nitrate solution, gives a black precipitate of metallic silver.

Phosphorus pentoxide is one of the most powerful dehydrating agents, and is used when it is desired to remove the last traces of aqueous vapour from gases. It will even dehydrate pure nitric and sulphuric acids, liberating the anhydride in each case, N_2O_5 and SO_3. When once hydrated, phosphorus pentoxide cannot again be liberated from the hydrate by any practicable method. When added to water a violent action takes place, but the solid does not dissolve very readily. A gelatinous mass is formed which only slowly passes into complete solution. The first and most energetic reaction is the production of metaphosphoric acid, HPO_3, and when this is once formed the dehydrating effect is greatly diminished:

$$P_2O_5 + H_2O = 2HPO_3.$$

Even when water is present in large excess further hydration is slow. In time, metaphosphoric acid passes into pyrophosphoric acid:

$$2HPO_3 + H_2O = H_4P_2O_7,$$

or, $$P_2O_5 + 2H_2O = H_4P_2O_7,$$

and finally into orthophosphoric acid:

$$P_2O_5 + 3H_2O = 2H_3PO_4.$$

On evaporating the solution these changes take place in reverse order. If heated until the temperature reaches approximately 140° C. and cooled, solid orthophosphoric acid is obtained in crystals which melt at 38·6° C. These are somewhat hygroscopic, and very soluble in water. On heating to a temperature of 300° C. the residue is pyrophosphoric acid, and this, heated to redness, leaves metaphosphoric acid, which resembles glass when cooled.

This, too, is hygroscopic and becomes sticky when exposed to air. In solution, as has been said, it passes over into orthophosphoric acid, slowly in the cold, and more quickly on boiling.

Phosphoric acid is usually made either from phosphorus pentoxide as described above, or by boiling phosphorus with moderately dilute nitric acid until oxidation is complete, and evaporating to crystallization. When required for the commercial preparation of its salts, it is obtained by boiling calcium phosphate with moderately dilute sulphuric acid and filtering off from the calcium sulphate. On adding sodium carbonate in the calculated quantity to this filtrate, sodium phosphate is obtained, which can be purified by recrystallization from water.

Very pure phosphoric acid can be obtained from sodium phosphate in the following way, which illustrates another general method, of wide application, for preparing acids from their salts (see p. 162).

To a solution of sodium phosphate, lead acetate solution is added, and the precipitated lead phosphate filtered off and thoroughly washed with water till everything soluble has been removed:

$$Na_2HPO_4 + Pb(CH_3COO)_2 = PbHPO_4 + 2CH_3COONa.$$

The white precipitate of lead phosphate is then washed into a flask with water, and hydrogen sulphide passed through the milk-like liquid for some time, after which it is boiled, and the black precipitate of lead sulphide filtered off:

$$PbHPO_4 + H_2S = PbS + H_3PO_4.$$

The filtrate is boiled until the odour of hydrogen sulphide is completely removed, and filtered again if necessary. The liquid can then be evaporated to any degree required.

Orthophosphoric acid forms three series of salts; primary, secondary, and tertiary (or normal), according as one, two, or all of the hydrogen atoms of the molecule are replaced by a metal or radical. The acid is therefore tribasic. The normal sodium salt, tertiary sodium phosphate, Na_3PO_4, is alkaline to indicators; the secondary, which is the salt generally called sodium phosphate, Na_2HPO_4, is neutral to litmus, and the primary, sodium dihydrogen phosphate, NaH_2PO_4, is acid. With the exception of the ammonium salts, the primary metallic salts on ignition usually leave metaphosphates, and the secondary, pyrophosphates. Microcosmic salt, the acid phosphate of sodium and ammonium, on ignition leaves sodium metaphosphate:

$$NaH_2PO_4 = NaPO_3 + H_2O,$$
$$2Na_2HPO_4 = Na_4P_2O_7 + H_2O,$$
$$Na(NH_4)HPO_4 = NaPO_3 + NH_3 + H_2O.$$

Since phosphoric acid is non-volatile, many metallic salts yield the corresponding acid when heated with it. Sulphuric acid, for instance, can be distilled from sodium sulphate by strongly heating a mixture of the salt and phosphoric acid:

$$Na_2SO_4 + 2H_3PO_4 = 2NaPO_3 + 2H_2O + H_2SO_4.$$

The superphosphate of lime, used in agriculture as a fertilizer, is a primary phosphate made by the action of the calculated quantity of chamber acid on calcium phosphate (apatite, coprolites, etc.):

$$Ca_3(PO_4)_2 + 2H_2SO_4 = 2CaSO_4 + CaH_4(PO_4)_2.$$

The ground-up phosphate and sulphuric acid are well mixed, and soon set to a hard mass, owing to the combination of the calcium sulphate (which is practically the same thing as plaster of Paris) with the water present in the acid. It is then reduced to powder, and used as a top dressing on the soil. Superphosphate, or acid phosphate, is comparatively soluble in water, but when moist it slowly reacts with the unchanged and insoluble tertiary salt $Ca_3(PO_4)_2$ which is always present, to form what is known as 'revert' or 'reverted' phosphate, secondary calcium phosphate, $CaHPO_4$. This latter compound, although almost insoluble in pure water, is slowly taken into solution by water containing carbon dioxide, and so becomes available as plant food.

Phosphorous acid is produced by the action of water on phosphorus trioxide:

$$P_2O_3 + 3H_2O = 2H_3PO_3,$$

but is more readily and conveniently made by the action of water on phosphorus trichloride, PCl_3 (*v. infra*); or by covering vitreous phosphorus with water, warming until the phosphorus is melted, and passing a slow stream of chlorine through the liquid until the phosphorus has disappeared:

$$PCl_3 + 3H_2O = H_3PO_3 + 3HCl.$$

As the acid is easily oxidized to phosphoric acid, it has strongly reducing properties. Salts of silver, gold, copper, and other metals yield the metal when warmed with a solution of phosphorous acid. The acid itself decomposes into phosphine, PH_3, and phosphoric acid on heating:

$$4H_3PO_3 = 3H_3PO_4 + PH_3.$$

The Halides of Phosphorus. When dry chlorine is passed over dry white phosphorus the energy of the combination usually causes the phosphorus to inflame. The first product of the union is the trichloride, PCl_3, which passes over into the pentachloride, PCl_5, in presence of excess of chlorine. These compounds are always prepared by the direct union of the elements, excess of phosphorus yielding the trichloride and excess of chlorine the pentachloride, and either variety of phosphorus may be used. The trichloride is purified from pentachloride by agitating with a little phosphorus, and afterwards distilling from the excess of phosphorus over a water bath, when it is obtained as a colourless liquid boiling at 76° C. The pentachloride is purified from trichloride by saturation with chlorine. It may also be obtained in colourless crystalline form by passing dry chlorine into a solution of phosphorus in carbon disulphide.

Phosphorus trichloride fumes strongly in moist air. It reacts with water to form hydrochloric and phosphorous acids, as already shewn. The phosphorous acid so obtained can be freed from hydrochloric acid by evaporating off the liquid at a moderate temperature.

Phosphorus pentachloride is ordinarily a yellowish rather hard solid which fumes strongly in moist air. The fumes are very irritating, rapidly destroying the mucous membrane of the nasal passages and producing effects resembling those of a cold. It is readily volatile and can only be melted by heating under pressure. Its vapour density, as determined in the ordinary way, is much less than the formula PCl_5 requires, at all temperatures. At 300° C. and above, it falls to about 52, which is half the calculated value. This is explained by the dissociation of the pentachloride into molecules of phosphorus trichloride and chlorine:

$$PCl_5 \rightleftharpoons PCl_3 + Cl_2.$$

This view is confirmed by determining the vapour density in an atmosphere of either phosphorus trichloride or chlorine, when its value is found to be increased.

Phosphorus pentachloride resembles the trichloride in its reaction with water. In presence of excess of water, the products of hydrolysis are orthophosphoric and hydrochloric acids:

$$PCl_5 + 4H_2O = H_3PO_4 + 5HCl.$$

If only a small quantity of water be used, an intermediate compound, phosphorus oxychloride, $POCl_3$, is produced:

$$PCl_5 + H_2O = POCl_3 + 2HCl,$$

which is, of course, further hydrolysed to orthophosphoric and hydrochloric acids by excess of water.

The oxychloride is easily prepared by the action of small quantities of water on phosphorus pentachloride, followed by distillation; or, more economically and conveniently, by distilling a mixture of equivalent quantities of pentachloride and pentoxide:

$$3PCl_5 + P_2O_5 = 5POCl_3.$$

Phosphorus oxychloride is a liquid boiling at 107° C., and resembles phosphorus trichloride in its general properties.

The replacement of hydroxyl groups by chlorine, by means of the penta- or tri-chloride of phosphorus, has already been mentioned. This is a very general reaction, since hydroxyl groups in very different states of combination can be so replaced. For instance, acids which contain this —OH group, on treatment with either of the chlorides of phosphorus, yield compounds of a new type, known as *acid chlorides*. This may be illustrated by the preparation of the acid chloride of acetic acid, that is, acetyl chloride. To anhydrous acetic acid contained in a distillation apparatus, the calculated weight of phosphorus pentachloride or trichloride is added, the mass being kept cool. When the addition is complete, the acetyl chloride, a liquid boiling at 55° C., can be distilled off from a water bath.

$$3CH_3COOH + PCl_3 = 3CH_3COCl + H_3PO_3.$$

Sulphuric acid reacts similarly with phosphorus penta-chloride, either one or both the hydroxyl groups being replaced by chlorine, according to the quantities used:

$$O_2S\diagup\diagdown_{OH}^{OH} \longrightarrow O_2S\diagup\diagdown_{Cl}^{OH} \longrightarrow O_2S\diagup\diagdown_{Cl}^{Cl}$$

sulphuric   chloro-sulphonic   sulphuryl
acid   acid   chloride

Acid chlorides are usually liquids. They fume in moist air, have an irritating effect on the eyes, and are characterized by three very general reactions.

1. With water, they yield the corresponding acid and hydrochloric acid:

$$O_2S\diagup\diagdown_{Cl}^{Cl} + 2H_2O = O_2S\diagup\diagdown_{OH}^{OH} + 2HCl.$$

2. With alcohols, they yield the corresponding *ester*, or ethereal salt, and hydrochloric acid:

$$CH_3COCl + C_2H_5OH = CH_3COOC_2H_5 + HCl.$$

3. They react with ammonia, the chlorine being replaced by an amino (NH_2—) group, to produce compounds known as *amides*. Thus, acetyl chloride yields acetamide:

$$CH_3COCl + 2NH_3 = CH_3CONH_2 + NH_4Cl.$$

Phosgene (p. 131) is the acid chloride of carbonic acid, although it cannot be prepared from the acid itself, for obvious reasons. It reacts similarly to the other acid chlorides, with water yielding hydrochloric and carbonic acids, the latter of course at once decomposing; with alcohol, diethyl carbonate is produced; with ammonia, the diamide of carbonic acid, known as carbamide, or urea, is formed:

$$OC\diagup\diagdown_{NH_2}^{NH_2} \longleftarrow OC\diagup\diagdown_{Cl}^{Cl} \longrightarrow OC\diagup\diagdown_{OC_2H_5}^{OC_2H_5}$$

Phosphine. When white phosphorus is boiled with caustic soda solution, a gas is evolved which inflames spontaneously when it comes in contact with air. Red phosphorus behaves similarly, if a concentrated solution of soda be used. The reaction may be represented:

$$4P + 3NaOH + 3H_2O = 3NaH_2PO_2 + PH_3.$$

Phosphine prepared in this way contains much free hydrogen, which is probably obtained from the hypophosphite by the further action of the soda:

$$NaH_2PO_2 + 2NaOH = Na_3PO_4 + 2H_2.$$

The spontaneous inflammability (with formation of smoke rings) is attributed to the presence of traces of liquid phosphine, P_2H_4. The presence of alcohol destroys this property, and when required in non-inflammable condition an alcoholic solution of potash or soda is used for the preparation. Spontaneously inflammable phosphine is prepared for technical purposes by the action of water on calcium phosphide, the latter being obtained by dropping pieces of phosphorus on to red hot quicklime in an inert atmosphere:

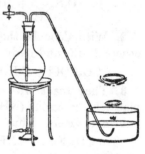

Fig. 33

$$Ca_3P_2 + 6H_2O = 3Ca(OH)_2 + 2PH_3.$$

Calcium phosphide is used in the Navy, under the name of Holmes's signal, for producing both flame and smoke by contact with water. A mixture of calcium phosphide and calcium carbide constitutes the so-called marine torch, which liberates both phosphine and acetylene, in a spontaneously inflammable condition, when thrown into water. The flame is a particularly brilliant one.

Phosphine is a poisonous gas having a disagreeable odour, and is but sparingly soluble in water. The aqueous solution is unstable, and when exposed to light the phosphine decomposes, depositing red phosphorus.

Phosphine resembles ammonia to some degree. It combines directly with halogen hydrides to form phosphonium halides, which are isomorphous with ammonium halides. These compounds are, however, very unstable:

$$PH_3 + HI \rightleftharpoons PH_4I.$$

The solubility of phosphine in water is not comparable with that of ammonia, nor is the aqueous solution alkaline to indicators. The ease with which phosphonium iodide decomposes renders it a convenient source of pure phosphine, which is obtained from it by the action of water, or better, by dilute alkalis. The function of the water or alkali is to remove the hydrogen iodide from the sphere of action, and thus to facilitate the decomposition.

$$PH_4I + KOH = KI + H_2O + PH_3.$$

CHAPTER VIII

ARSENIC AND ITS COMPOUNDS

ARSENIC is found naturally, to a small extent, in the uncombined condition, but the chief sources of it are the metallic sulphide ores, very few of which are quite free from arsenic. Since the presence of this element has an injurious effect on most metals, it is removed from the ores before smelting, and much of the arsenic in the market may be regarded as a bye-product from metallurgical works. Many metallic oxide ores also contain it, notably the tin ores of Cornwall. Two mineral sulphides of arsenic are fairly abundant, red orpiment or realgar, As_2S_2, and yellow orpiment, As_2S_3, but the ore mainly worked for arsenic as the principal product is mispickel or arsenical pyrites, $FeS_2 . FeAs_2$.

The extraction of arsenic from its ores is a very simple process, whether the object is the removal of traces of arsenic as an undesirable impurity, or the commercial winning of the element. Arsenic itself, or any compound containing it, on being heated strongly in a current of air yields the volatile oxide As_2O_3, which can be condensed by cooling. The operation is carried out on the hearth of a reverberatory furnace, where the ore is heated to redness in a current of air. By using a revolving furnace the process can be made continuous, the ore being fed in at the higher end and the spent material discharged at the lower. The fumes pass through a series of brickwork or earthenware chambers, where condensation of the arsenic trioxide takes place before the waste gases are allowed to escape.

From the white arsenic, As_2O_3, so obtained, the element can be prepared by a process of reduction. An intimate mixture of powdered charcoal and arsenic trioxide is gently heated, when the element sublimes and condenses on the colder surfaces of the apparatus, in the form of dark grey lustrous crystals, which present a mirror-like appearance when deposited upon glass. It is a brittle solid which sublimes without melting unless heated under pressure; is insoluble in water and in acids generally, though nitric acid oxidizes it to the soluble arsenic acid. The element itself has but little application. The presence of traces of arsenic in copper enormously reduces its conductance for electricity, but renders the copper more resistant to the action of the products of combustion of coal. For this reason, copper containing a little arsenic is sometimes used for plates and tubes in steam boilers. A trace of arsenic is often present in speculum metal, where it is thought to improve the reflecting surface. Small quantities of arsenic are usually added to the lead for making small shot, with the idea of increasing the surface tension of the molten metal and so ensuring more spherical shot, and also somewhat hardening the metal.

White arsenic sublimes without melting, and is purified in this way. If the vapour be condensed on a surface just cold enough for the purpose, it separates as a coherent solid which shews a conchoidal fracture resembling that of broken glass. This is the amorphous or vitreous modification. The clear transparent masses soon acquire a porcelain-like appearance and texture, extending inwards from the surface, due to slow crystallization. By rapid cooling of the vapour a crystalline form is produced, for the greater part octahedral. A monoclinic form has also been described. Both crystalline forms are more stable than the vitreous. All forms dissolve slightly in water, the solution having feebly acidic properties. Concentrated hydrochloric acid reacts with arsenious anhydride forming the chloride, which while soluble in concentrated hydrochloric acid, undergoes hydrolysis on dilution. The crystals of arsenious anhydride separate from solution with the production of flashes of light.

$$As_2O_3 + 6HCl \rightleftharpoons 2AsCl_3 + 3H_2O.$$

White arsenic is very poisonous, and being tasteless, is on this account the more dangerous. The fatal dose is stated to be 60 milligrammes, but it greatly depends on the physical condition of the person taking it. Much smaller quantities have been known to cause death, while, on the other hand, the frequent use of small doses so inures the system to the poison that comparatively large quantities may be taken without fatal effects. Cases of poisoning have been observed as the result of eating apples from orchards where the trees were sprayed in the early summer with a dilute solution of arsenic, the object being the destruction of parasites. The quantities of arsenic taken in these cases must have been very minute.

The feebly acidic solution obtained by saturating water with arsenious oxide probably contains arsenious acid, but this compound has not been isolated. Metallic salts can be prepared from the solution by exactly neutralizing it with ammonia, and adding a solution of silver nitrate or copper sulphate, etc., the arsenites of the metals being thereupon precipitated. These precipitated salts correspond to a tribasic acid of the formula H_3AsO_3. The copper salt, known as Scheele's green, to which the formula $CuHAsO_3$ is usually assigned, is used as a pigment. The sodium and potassium salts, which are soluble, are used in sheep dipping preparations as insecticides, for skin and leather dressing and preserving, and as weed killers.

Nitric acid oxidizes arsenic trioxide to arsenic acid:

$$2As_2O_3 + 4HNO_3 + 4H_2O = 4H_3AsO_4 + 2N_2O_3.$$

White arsenic, boiled with moderately concentrated nitric

acid and evaporated down, on cooling deposits hydrated crystals of arsenic acid. Like orthophosphoric acid, arsenic acid loses water on heating, and at 180° C. passes into pyroarsenic acid, $H_4As_2O_7$. At about 200° C. more water is lost, and metarsenic acid, $HAsO_3$, is formed. Unlike metaphosphoric acid, which volatilizes unchanged at a white heat, metarsenic acid yields its anhydride, arsenic pentoxide, As_2O_5, in the form of a glass-like solid, when heated to incipient redness. At a red heat the pentoxide decomposes into trioxide and oxygen.

The salts of ortho-, pyro- and metarsenic acids are isomorphous with the salts of the corresponding acids of phosphorus, but are less stable.

Arsine, AsH_3, is the only compound of arsenic with hydrogen the existence of which is beyond doubt. It cannot be made by the *direct* action of hydrogen on the element, like the corresponding hydride of nitrogen, but practically any compound of arsenic will yield arsine in presence of 'nascent' hydrogen, and this reaction is the basis of the best known tests for the presence of arsenic. In chemico-legal cases the one most commonly employed is known as Marsh's test. This depends upon the formation of arsine by the union of arsenic and hydrogen, the latter being generated electrolytically, or by the action of zinc upon dilute hydrochloric or sulphuric acid. An apparatus for generating hydrogen is set up, and after the air has been swept out, the hydrogen is lighted at a jet. It is necessary to ascertain that the zinc and acid are themselves free from arsenic before proceeding further, by making what is known as a blank test. A piece of cold porcelain, such as a crucible lid, is brought into the flame of the burning hydrogen for a few seconds. If no brown stain is produced on the porcelain, the materials are free from arsenic and the test may proceed. A few drops of the solution to be examined are introduced into the apparatus. If arsenic be present, a change in the appearance of the flame is observed; the almost colourless flame of pure hydrogen changing to a bluish-violet. When a cold porcelain surface is held in the flame, a brown mirror-like deposit of elementary arsenic is produced, which is readily soluble in solutions of hypochlorites, such as bleaching powder, or sodium hypochlorite.

If arsine, carefully freed from traces of hydrogen sulphide by passage through a wash bottle containing a solution of lead acetate, be passed into a solution of silver nitrate, a precipitate of metallic silver is produced, while arsenious acid remains in solution:

$$AsH_3 + 6AgNO_3 + 3H_2O = H_3AsO_3 + 6HNO_3 + 6Ag.$$

Antimony compounds behave in a somewhat similar manner, but there is little risk of confusion because (1) the hydrogen flame containing antimony hydride (stibine, SbH_3) has a greenish colour, (2) the antimony mirror is practically insoluble in hypochlorites, and (3) when passed into silver nitrate solution, the gas yields a black precipitate of silver antimonide, $SbAg_3$, but no acid corresponding to arsenious acid is found in the solution.

$$SbH_3 + 3AgNO_3 = SbAg_3 + 3HNO_3.$$

Another test, due to Fleitmann, consists in generating hydrogen by the action of aluminium on sodium hydroxide, adding a little of the substance to be tested, and allowing the gas evolved to react with silver nitrate. This can be done by holding a piece of filter paper, moistened with a drop of silver nitrate solution, over the mouth of the test tube. If arsine be present, a black stain of metallic silver appears on the paper. As antimony hydride is not produced in alkaline solutions, there is no possibility of mistaking between antimony and arsenic when Fleitmann's method is used, and furthermore, the gas cannot contain hydrogen sulphide when generated from alkalis.

A third test for arsenic is that devised by Reinsch. A clean strip of copper foil of high purity is placed in the solution to be tested, and hydrochloric acid added. If arsenic be present, a deposit of arsenic (or copper arsenide) appears on the surface of the metal on warming, mirror-like if thin, but if much arsenic be present, of a dull grey appearance. The strips of arsenic-covered copper can be removed, washed with water, dried, and thereupon heated in a clean dry test tube, when a crystalline deposit of arsenious oxide will be formed in the upper and colder parts of the tube. These crystals can be examined with the aid of a lens.

In chemico-legal cases it is customary to pass the gas issuing from Marsh's test slowly through a glass tube heated to low redness. The arsine decomposes under moderate heating, and the arsenic adheres to the glass. The quantity of arsenic present is estimated by the appearance of the mirror in the glass tube, which is compared with a series of standard tubes similarly prepared from known weights of arsenic.

Pure arsine is rarely prepared except for special experimental purposes, when it can be obtained by fractionating the gaseous product of the action of dilute acids on zinc arsenide. It is exceedingly poisonous, very sparingly soluble in water, and has no basic reaction towards indicators. As has been seen, arsine burns to arsenious oxide and water in a full supply of air, while water and arsenic are formed when the supply of oxygen is limited.

The sulphide of arsenic, As_2S_3, may be prepared by the action of hydrogen sulphide on arsenites in presence of dilute acids. Arsenic acid and arsenates are slowly reduced, in the first place to arsenites, by hydrogen sulphide; sulphur being precipitated:

$$H_3AsO_4 + H_2S = H_3AsO_3 + H_2O + S.$$

On further passage of the gas, yellow arsenious sulphide is precipitated:

$$2H_3AsO_3 + 3H_2S = As_2S_3 + 6H_2O.$$

CHAPTER IX

BORON, SILICON AND COLLOIDS

BORON in the elementary state is of little importance, and is not so found in nature. It occurs in combination as borax and as boric acid in many parts of the world. For long, native borax or 'tincal' was imported from Tibet and India, and more recently considerable supplies have come from lacustrine deposits in California. Much of the borax of commerce is now derived from the mineral boro-natro-calcite imported from Peru. In the volcanic districts of Tuscany steam jets (soffioni) issue from the ground, carrying with them traces of boric acid, and this accumulates in artificial lagunes built round the soffioni and through which the steam is made to pass. The water from the lagunes is evaporated by the steam from the soffioni themselves, which passes under the evaporating pans before entering the lagunes. This system has now been in operation for over a century, and considerable quantities of boric acid are obtained by it.

Borax is the hydrated sodium salt of pyroboric acid, $Na_2B_4O_7 . 10H_2O$. It is very soluble in water, to which it gives an alkaline reaction. If concentrated hydrochloric acid be added to a saturated solution of borax, orthoboric acid, H_3BO_3, is precipitated, and can be filtered off and washed free from sodium chloride by cold water. On heating at 100° C., this acid loses water and passes into metaboric acid:

$$H_3BO_3 = HBO_2 + H_2O.$$

At 140° C. more water is lost and pyroboric acid formed:

$$4HBO_2 = H_2B_4O_7 + H_2O,$$

while at red heat the pyroboric acid loses all its hydrogen in the form of water, and leaves the oxide, B_2O_3, as a fused glassy mass which slowly volatilizes at high temperatures. In presence of water the oxide is again hydrated to ortho-boric acid.

When borax is heated it intumesces, becoming a bulky opaque mass, with loss of both water and boric acid:

$$Na_2B_4O_7 . 10H_2O = 2NaBO_2 + 2H_3BO_3 + 7H_2O.$$

Ultimately it fuses into a glass-like mass of sodium meta-borate (the borax bead used in qualitative analysis). This, in a fused condition, dissolves many metallic oxides, and these often impart to it a colour characteristic of the metal, by which the latter can be identified. The use of borax as a flux in soldering and brazing depends on this property of dissolving metallic oxides. Borax is also used in the manu-facture of special kinds of glass, in laundry work for glazing linen, and in soap making.

Orthoboric acid is a pearly crystalline solid, slightly volatile in steam, and sparingly soluble in cold water. The solution has a faintly acid reaction, and when neutralized with sodium hydroxide forms sodium metaborate:

$$NaOH + H_3BO_3 = NaBO_2 + 2H_2O.$$

Boric acid, or boracic acid as it is still sometimes called, is a mild antiseptic, and on this account has been used as a preservative for cream and other foods. It is employed in surgical dressings, such as lint, antiseptic cotton wool, etc., and in eye lotions.

Basicity of Acids. The basicity of an acid usually means the number of kinds of salts which the acid can form with a univalent metal such as sodium or potassium. For instance, hydrochloric acid and nitric acid each form one, and only one, such salt. They are therefore monobasic.

Sulphuric acid forms two potassium salts, the normal sulphate, K_2SO_4, and the acid sulphate, $KHSO_4$, and is therefore dibasic. Similarly phosphoric acid, H_3PO_4, is tribasic. Basicity is not defined by the number of hydrogen atoms in the molecule, but by the number of kinds of potassium or sodium salts the acid can form as stated above, or, by the number of hydrogen atoms in the molecule replaceable by the metal. An acid may contain hydrogen atoms which are not replaceable in this way. Acetic acid, for instance, which has the molecular formula $C_2H_4O_2$, is a monobasic acid, since but one sodium acetate is known, and only one of the hydrogen atoms in the molecule is replaceable by a metal. The formula is on this account usually written $C_2H_3O_2H$, or better, CH_3COOH, in order to differentiate between the acidic and the non-acidic hydrogen atoms. The three non-acidic hydrogen atoms are easily replaced by chlorine atoms, but not by metals.

It is not possible to prepare three sodium or potassium salts from orthoboric acid. The only salt obtainable by the interaction of solutions of caustic soda and this acid is sodium metaborate, $NaBO_2$. The difficulty is increased by the fact that the sodium salt contains no hydrogen. This consideration by itself would indicate the formula $HBO_2 . H_2O$ rather than $B(OH)_3$. As judged by both the definitions given above, orthoboric acid is *mono*basic.

Silicon is sometimes said to be the most abundant element in nature after oxygen. This statement is based upon the probability that the interior of the earth does not differ greatly in composition from the lavas extruded from volcanoes, which contain more than half their weight of silica, SiO_2. The element itself is of little importance. The only oxide of silicon is SiO_2, which occurs widely distributed as sandstone; in a granular condition as sand; as a constituent of many rocks; as flint; and in a very pure form as quartz. Sand when pure is of a dazzling whiteness, as may be seen in that of Alum Bay (Isle of Wight), Morar in Inverness-shire, Bohemia, and elsewhere. Usually it is coloured with metallic oxides, the commonest of which is ferric oxide. Fused silica is replacing glass in the manufacture of many kinds of apparatus for scientific and technical use. Although this material cannot be used in presence of alkalis, by which it is readily attacked, it has many advantages over glass. It is not acted upon by any acid, except hydrofluoric. Its coefficient of expansion is so small that sudden changes of temperature do not injure it. For instance, a red hot crucible of silica may be plunged into cold water without breaking. The fused material can be drawn into almost microscopically thin fibres, which are valuable for torsion measurements owing to their remarkable elasticity.

Silica is an acidic oxide. It does not 'unite' (*vide infra*) with water directly, but can be made to do so by indirect methods. When fused with sodium carbonate, carbon dioxide is liberated, a reaction common to all acidic oxides. The constitution of the derivatives of silica is complicated. Silicates are better understood by writing them on Berzelius's dualistic system, which though now obsolete, is useful here. On this system, salts are represented as compounds of basic oxides with acidic oxides. Sodium sulphate, for instance, becomes $Na_2O . SO_3$, and the silicates of sodium $(Na_2O)_x . (SiO_2)_y$. As usually prepared, so-called sodium silicate is not a chemical individual but a mixture of compounds, each of which can be represented by the above formula. When y is comparatively large the sodium silicate is insoluble, non-crystalline, and transparent. Ordinary glass is mainly such a mixture, though other silicates are always present. When x is large the silicate is soluble, the material being known as soluble glass, or water glass, and is used as a preservative for eggs, and stone in buildings and monuments.

It must not be supposed that compounds represented in this rather indefinite way depart from the law of constant proportions. They are mixtures of quite definite compounds, but in most cases mixtures whose constituents are difficult to separate and isolate. The more important silicates found in nature are those of the alkali metals together with aluminium or magnesium. Clay, for example, is chiefly aluminium silicate, kaolin or colourless China clay having the formula $Al_2O_3 . 2SiO_2 . 2H_2O$. Asbestos, meerschaum, and French chalk or talc, are silicates of magnesium. Felspar is a double silicate of potassium and aluminium,

$$K_2O . Al_2O_3 . 6SiO_2.$$

When a dilute solution of water glass is acidified with very dilute hydrochloric acid, and the whole dialysed (*vide infra*) until all chlorides are removed, a solution of silicic acid remains as a clear transparent colourless liquid. On standing, a cloudiness or jelly develops, which thickens and shrinks away from the water, in course of time leaving a horn-like deposit, small in bulk, which cannot be re-dissolved. If this be removed and heated until the weight remains constant, the residue consists of pure silica. This is the basis of gravimetric estimations of silica in combination. For instance, the estimation of silica in glass is carried out by fusing a weighed quantity of the finely powdered glass with at least six times its weight of anhydrous sodium carbonate, in a platinum crucible over the flame of a blast lamp, until all evolution of carbon dioxide ceases. When cold, the contents of the crucible are extracted with water and the solution acidified with hydrochloric acid, and then evaporated to dryness on a water bath. The solid residue contains nothing insoluble in hot

dilute hydrochloric acid except silica. It is repeatedly extracted with this reagent until all soluble material has been
removed, and thereupon ignited and weighed as SiO_2.

Although formulae have been assigned to certain silicic
acids such as H_4SiO_4 and H_2SiO_3, it has been shewn by
van Bemmelen that water associated with silica is to be
regarded as adsorbed rather than in chemical combination.
This view is supported by the fact that no definite acids
have been isolated. For instance, when a dilute solution
of so-called silicic acid is evaporated, there is no point at
which the vapour pressure suddenly changes, as would be
the case if a definite compound existed.

Glass. As has been stated above, ordinary glass is a
mixture of silicates, in which those of sodium usually
predominate, though the composition can be varied in
many ways. A dense kind, with a high refractive index and
low fusibility, contains silicate of lead. There are many
formulae in use for the preparation of optical glass, and
much attention has been given of late to this highly
technical manufacture. The material in use for boiler
gauges and thermometers contains boro-silicates, as the
replacement of some of the silica by boric anhydride seems
to render the material tougher and more resistant to the
action of hot liquids.

As is well known, glass softens gradually when heated,
and so can be worked into any desired shape. It should
afterwards be cooled very slowly and carefully, a process
known as annealing. Badly annealed glass is very liable
to fracture, especially on heating, since the exterior has
been cooled more quickly than the interior, and the whole
mass is under considerable strain.

Fused glass, like fused borax, can dissolve many metallic
oxides, and it is in this way that coloured glasses are obtained. The pale green tint of ordinary window glass is due

to ferrous oxide, while the yellow 'non-actinic' kind contains ferric oxide. Ruby glass is coloured with cuprous oxide.

In course of time glass frequently assumes a crystalline form, with loss of transparency, or becomes 'devitrified.' This can be seen sometimes in ancient window panes, and the same phenomenon may be observed when old glass tubing is heated to low redness.

Water has an appreciably solvent action upon some kinds of glass. The action consists essentially in hydrolysis of the alkaline silicate, and new surfaces are always more corroded than those which have been previously exposed to the action of water. As might be expected, alkaline solutions exert greater solvent action than does water. The only acid which will dissolve glass is hydrofluoric. Etching is carried out by coating the glass surface with a thin film of molten paraffin wax. When this solidifies, it can be scratched off with the point of a needle in the desired pattern, and the surface thus uncovered exposed to the action of either hydrofluoric acid or hydrogen fluoride. After the action has proceeded far enough, the acid is washed away with water and the rest of the wax removed with either petrol or benzene. Etched marks do not render the glass so liable to fracture as do the scratches made with a diamond.

The slag which runs in a molten condition from the blast furnaces of iron works is in all essentials a glass. The earthy matter accompanying the ore is often highly siliceous, and the material added to react with it in order to produce something which will flow is, in such cases, limestone. These two combine to produce an easily fusible silicate of calcium. Slags of this nature are sometimes ground to powder, and sold as hydraulic mortar. The 'setting' of such is due to slow union with water to form a crystalline but insoluble mass, which is not further acted on by water. Portland cement is a very similar material, being essentially a silicate of calcium, and owes its hardening property to the same cause, but good Portland cement has never been actually fused, and therefore grinds to a much finer powder. In consequence it 'sets' very much better. Portland cement is so called because it resembles Portland stone when set.

Colloids. In the middle of the nineteenth century Graham carried out many experiments on the diffusion of substances in solution. One of his methods consisted in immersing an open bottle, filled with a concentrated solution of the substance under investigation, in a large trough of water, care being taken to avoid mixing by convection. At definite intervals of time, portions of the liquid were removed from a fixed part of the trough and subjected to analysis. In this way the amount of solute which diffused from the concentrated solution in the bottle to the fixed point in the trough was determined. Graham found that substances such as sugar, acids, bases, and salts, diffused readily, but that others, such as gelatine, albumin, and silicic acid, scarcely diffused at all. The former class he termed *crystalloids*, since they were for the greater part substances of a crystalline nature, and the latter class *colloids*, from their general glue-like characteristics. This diffusibility of a crystalloid such as copper sulphate is shewn as a laboratory experiment by filling a tall cylinder with water, and then introducing into the lower part of it a saturated solution of copper sulphate, in such a manner as to avoid mixing as far as possible. This is done by means of a thistle funnel, ending in a fine aperture, which reaches to the bottom of the cylinder. With care, the copper sulphate solution can be introduced in such a way that there is a sharp line of demarcation between it and the water.

If the vessel be kept under such conditions that it is not shaken or otherwise disturbed, the copper sulphate diffuses upwards into the water, the line of demarcation becoming more and more vague. Weeks may elapse, however, before copper sulphate reaches the upper part of the cylinder. Of course the final condition of the liquid is that a homogeneous solution is formed, which shews no tendency to separate again.

Experiments on the separation of crystalloids and colloids, based on the diffusibility of the former and the non-diffusibility of the latter, were carried out later. Graham found that parchment paper allows crystalloids to pass through it, but that colloids are retained by such a membrane. Thus, a solution of colloidal silicic acid can be prepared by allowing the products obtained by fusing silica with sodium carbonate, extracting with water and then acidifying with dilute hydrochloric acid, to diffuse through a parchment membrane, such as is shewn in Fig. 34, floated on water which is frequently changed. Everything in solution ultimately passes through the membrane, except the hydrated silica. Such a membrane can be made by introducing a piece of filter paper into a cold mixture of one volume

Fig. 34

of concentrated sulphuric acid and two volumes of water. After a few seconds immersion, the paper is transferred to water and well washed. Under this treatment the paper becomes much tougher, and closer in texture.

This process was termed *dialysis* by Graham, and was successfully applied by him to problems in toxicology. Difficulties frequently arise in testing the contents of a stomach for poisons, owing to the presence of organic colloidal matter. When, however, these are subjected to dialysis, the inorganic crystalloid poisonous material, such as corrosive sublimate or white arsenic, diffuses through the dialysing membrane into the external water. After concentration of the latter by evaporation, it can be examined without difficulty.

Graham seems to have regarded crystalloids and colloids as essentially different classes of substances, but there is no sharp line of division between them as regards diffusibility. All crystalloids do not diffuse at the same rate, and, on the other hand, all substances which from their general characteristics would be regarded as colloidal, do not utterly refuse to pass through a dialysing membrane. Indeed it is possible in some cases to effect at least a partial separation between colloids themselves by the process of dialysis. The phenomenon is one of degree, and a very important factor is the size, or diameter, of the colloidal particles. Although the latter are very large as compared with, say, a molecule of water, yet in colloidal solutions, or 'sols,' they are far below the *minimum visibile* of the microscope in the usual way. By an adaptation of the ordinary microscope, known as the ultra-microscope, it is possible to see the scattering effect upon light which these tiny particles produce, and so to demonstrate their existence, but nothing resembling an image of the particle itself is produced on the retina.

Colloidal solutions are intermediate in properties between true solutions and suspensions of insoluble substances. They are sometimes called disperse systems, the minute particles being dispersed in a liquid, the 'dispersion medium.' A distinction is sometimes drawn between 'suspensoids,' which are solids dispersed in a liquid, and 'emulsoids,' which consist of two immiscible liquids, one dispersed in the other, such as milk.

It is now recognized that many substances formerly regarded as insoluble can be obtained by suitable means in colloidal solution, such as sulphur, platinum, gold and other metals, ferric and other hydroxides, and many metallic sulphides. The method of electric dispersion devised by Bredig is of general application for the preparation of colloidal solutions of metals. This consists in establishing an electric arc between electrodes of the metal, below the surface of water. A colloidal solution of platinum prepared in this way, containing only about one milligramme of metal per litre, is a dark brownish-violet liquid possessed of very active catalytic properties. Gold sols have a violet to purple colour, and seem to have been known as early as the middle of the seventeenth century, when they were credited with extraordinary curative properties. Some of those prepared by Faraday in 1857, by the action of reducing agents on solutions of gold salts, still shew faint colours, though the gold from the rest of them has separated out.

By subjecting a solution of ferric chloride to dialysis, preferably in running water, until the contents of the dialyser give no reaction with silver nitrate, a colloidal solution of ferric hydroxide can be obtained. The chloride hydrolyses in aqueous solution thus:

$$FeCl_3 + 3H_2O \rightleftharpoons Fe(OH)_3 + 3HCl,$$

and as the hydrochloric acid is rapidly removed from the sphere of action by dialysis, the reaction completes itself from left to right with the result stated.

If hydrogen sulphide be passed through a solution of arsenious oxide in water, the liquid turns bright yellow from the formation of a colloidal solution of arsenious sulphide, but no precipitate appears unless the liquid is acidified. This solution may be freed from dissolved hydrogen sulphide by aspirating an inert gas through it, and is then remarkably stable. A similar solution of colloidal antimony sulphide, having a dark red colour, may be prepared in the same way. These are *negative* colloids; that is, the particles can be shewn to possess a negative charge, by placing two electrodes in the liquid and applying a suitable difference of potential, when the particles will be found to wander towards the positive pole. Ferric hydroxide as prepared above, on the other hand, is a positive colloid, since it drifts towards the negative electrode.

Addition of electrolytes to colloidal solutions will cause coagulation, but some salts are much more efficient in this respect than others. When colloidal solutions, the particles of which have charges of the same sign, are mixed, pre-cipitation does not take place. For example, colloidal solutions of arsenious and antimonious sulphides may be mixed without apparent change. On the other hand, when colloidal solutions of opposite charges come in contact, the charges are neutralized and a precipitate separates. Thus if a colloidal solution of ferric hydroxide be added to one of either arsenic or of antimony sulphide, a precipitate at once appears. The presence of an electric charge seems to be necessary for the stability of colloidal solutions. Col-loids are becoming of increasing importance in many directions. Many medicinal preparations contain the active material in colloidal form, since as such it seems to be more easily assimilated.

CHAPTER X

FURTHER PROPERTIES OF SOLUTIONS.
THE IONIC THEORY

Osmosis. If the shell be dissolved from an egg by dilute
hydrochloric acid, and the egg be then placed in distilled
water, in the course of one or two days it will increase considerably in size. It can have gained nothing but water.
A similarly prepared egg placed in brine, shrinks. It can be
shewn to have lost nothing but water. If the shrunken egg
be placed in distilled water, and the swollen egg in brine,
the processes are reversed. Plainly then the outer membrane of eggs permits of the passage of water and water
only. Such a membrane is 'semi-permeable.' Turnips may
be used to shew the same phenomenon; when placed in
water they swell, in brine they shrink, the process, as before, being reversible.

The phenomenon can be studied more closely in the
following way. A long thistle funnel has the wide end
covered with bladder skin firmly bound on. It is filled to
a short distance up the stem with a solution of cane sugar,
the filled end being immersed in a beaker of water, and the
funnel clamped in a vertical position, in such a way that
the sugar solution within and the water without are on the
same level. In course of time the liquid rises in the stem,
thus developing a pressure which can be measured. Ultimately the liquid ceases to rise, the passage of water into
the sugar solution being stopped by the pressure (Fig. 35).

Perfect semi-permeable membranes, that is, those which
permit of the passage of the solvent only and not of the
solute, do not exist, but bladder skin is a fairly good one.
A better membrane may be made by blocking up the

minute interstices of a small porous earthenware pot, about
8 cm. in length and with an internal diameter of about

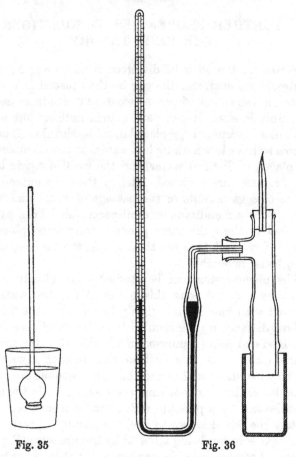

Fig. 35   Fig. 36

2 cm., with copper ferrocyanide or other similar compound.
This is done by completely filling the vessel with a solution
of potassium ferrocyanide, and immersing in a solution of

copper sulphate. The two solutions gradually permeate the porous walls of the cell, and on meeting in the interior of the wall, produce a precipitate of copper ferrocyanide, which closes the pores so effectively as to render the cell almost semi-permeable in the sense defined. The cell thus prepared should be washed very thoroughly with water before being taken into use. Such a vessel is filled with the solution to be examined, fitted with a rubber stopper carrying a manometer tube, and the joints strengthened and protected with a thick coat of sealing wax or other similar material. This apparatus is sufficiently perfect to give good pressure measurements in the case of dilute solutions (Fig. 36).

A series of experiments with such an apparatus will shew:

1. The pressure developed by an aqueous solution of cane sugar is proportional to the concentration of the solution, at a given temperature.

2. For solutions of a given concentration, the pressure increases with rise of temperature in accordance with the law of Charles for the expansion of gases; that is, the pressure is directly proportional to the temperature on the absolute scale.

3. The pressure developed by a given solution at any temperature is precisely that which would be caused if the mass of solute present existed in the form of a gas, having the same volume and temperature as the containing vessel, or, in other words, one gramme molecule of solute contained in one litre of solution would have an 'osmotic' pressure of 22·2 atmospheres at 0° C.

If similar experiments be made with glucose or urea as solute, the same numerical relationships will be found. The molecular weight of cane sugar, $C_{12}H_{22}O_{11}$, is 342; that of glucose, $C_6H_{12}O_6$, is 180; that of urea, CON_2H_4, is 60.

Solutions of cane sugar, glucose, and urea, containing
respectively 342 grm., 180 grm., and 60 grm. of each of
these compounds, each dissolved in, say, 10 litres of solution, would give the same osmotic pressure at the same
temperature. Solutions having the same osmotic pressure
are termed *isotonic*. It follows that equimolecular solutions
of this kind are isotonic, and also that these experiments
present a method of comparing (and therefore of determining) the molecular weights of such substances *in solution*.

Although a *perfect* semi-permeable membrane seems to
be incapable of experimental realization, it can nevertheless
be shewn that the osmotic pressure of a solution does not
depend on the nature of the membrane. Imagine a glass
tube, closed at each end with such a membrane and filled
with a solution. Assume also that one of the membranes
A is capable of giving rise to an osmotic pressure P, while
the value of the osmotic pressure developed by the other
membrane B is p, P being greater than p. If such an apparatus be immersed in the pure solvent, the latter can
effect an entry by both membranes. This process would
continue until the maximum value of the osmotic pressure
capable of being given by the membrane B is reached,
viz. the value p. At this stage the solvent would cease to
enter the solution by the membrane B, but as the membrane A is capable of realizing a higher osmotic pressure,
the solvent would continue to enter by that membrane, and
would pass out from the solution through the membrane
B, in consequence of the difference of pressure, $P - p$.
Clearly this process could continue indefinitely, and therefore be capable of doing work. In other words, the system
would constitute a *perpetuum mobile*. Since such a state of
affairs is contrary to experience, the conclusion is inevitable
that $P - p = 0$. That is, the values of the osmotic pressures
produced by the two membranes must be identical.

Owing to experimental difficulties, determinations of molecular weights in solution are rarely made by the direct measurement of osmotic pressure. It has been shewn however by van 't Hoff that certain other properties of dilute solutions, particularly the depression of the freezing point and the elevation of the boiling point, both of which are easily measurable, are directly proportional to the osmotic pressure. Methods for determining molecular weights, based on these properties, are in general use.

The separation of the solute when a concentrated solution is cooled has been described in Chap. II. The effect of cooling very dilute solutions is to crystallize out the pure solvent. This can be shewn by freezing a dilute solution of potassium permanganate, when the highly coloured liquid deposits colourless ice. Since the cooling of concentrated solutions leads to a separation of solute, while freezing dilute solutions results in the separation of pure solvent, it would be reasonable to expect that solutions could be prepared of such concentration as to deposit *both* solute *and* solvent.

Suppose, for instance, two solutions of common salt, one saturated and the other very dilute, are cooled. In the one case common salt crystallizes out; in the other pure ice separates. If the crystals be continuously removed as they are formed, the first solution becomes progressively more and more dilute, while the other becomes more concentrated. Clearly a point would be reached at which the concentration of the two solutions would be the same. Further cooling would result in the separation of crystals containing both salt and ice, and these would have the same composition as the remaining liquid. At this point the solution solidifies as a whole. The crystals consist of 23·6 per cent. of sodium chloride and melt at − 23° C. This was formerly considered to be a definite compound of sodium chloride and water, to which the name *cryohydrate* was given. It is now known not to be a compound but a mixture, having a constant freezing point, analogous to the constant boiling mixtures previously described. Such constant freezing mixtures are now termed *eutectics*.

Electrolytes. Aqueous solutions of sodium chloride differ as regards their osmotic effects from solutions of substances such as cane sugar. The value of the osmotic pressures (or other properties proportional to these, such as depression of the freezing point, and elevation of the boiling point) of dilute solutions of sodium chloride approximate to *double* that calculated from the formula. Substances may therefore be divided into two classes, namely, those of which the molecular weights in solution are in agreement with the values obtained from other considerations, and others, such as sodium chloride, which give rise to abnormally high osmotic effects, thus indicating a *lower* value for the molecular weight. This classification is further justified by the fact that the latter substances, in aqueous solution, are electrolytes, while the former are non-electrolytes. Practically all salts, and most acids and bases, belong to this latter class.

When a solution of dilute sulphuric acid is electrolysed, the gases appear at the electrodes only, not in the space between them, hydrogen being evolved at the cathode and oxygen at the anode. The question as to *where* the water molecule is decomposed presents some difficulty. As only one gas is evolved at each electrode, decomposition cannot take place on the electrodes themselves. Nor can the process take place in the intervening space, because the evolution of the gases is confined to the electrodes. No theory of electrolysis can be considered as satisfactory which fails to account for these familiar phenomena.

The quantitative laws of electrolysis, which were established by Faraday in 1833, may be stated as follows:

1. The amount of electrolytic decomposition which takes place is directly proportional to the quantity of electricity which passes through the circuit, that is, to the strength of the current, and to the time.

2. When the same current is passed through a series of different electrolytes, the masses of the several substances liberated at the electrodes are proportional to their chemical equivalents.

Electrolytes obey Ohm's law, viz. that the current flowing through the circuit is directly proportional to the applied electromotive force, and inversely proportional to the resistance; in this respect they resemble conductors of the first class. This can be shewn by the use of two copper electrodes immersed in a solution of copper sulphate, in circuit with a suitable source of current and a galvanometer, and provided with an arrangement for measuring the difference of potential across the electrodes. Any difference of potential, however small, applied to the electrodes, will result in a current flowing through the system, and the strength of the current as indicated by the galvanometer will be strictly proportional to this electromotive force. No gas is evolved at the electrodes immersed in the solution of copper sulphate, but the passage of the current is none the less accompanied by electrolysis, because the anode gradually goes into solution, and at the same time copper is deposited from the solution on the surface of the cathode, as may easily be verified by weighing the two electrodes before and after the experiment. As electrolytic conduction takes place on the application of very small differences of potential, the process must be considered primarily as involving a transfer of matter rather than decomposition. If an electrolytic cell, consisting of two platinum electrodes immersed in dilute sulphuric acid, be substituted for the copper voltameter in the above experiment, the phenomena are not quite so simple. If the difference of potential applied to the electrodes is below a certain value, approximately 1·7 volts, no gas is evolved at the electrodes, nor does any appreciable current flow through the system. When the difference of potential exceeds this critical value, continuous electrolysis takes place and Ohm's law is once more obeyed. This phenomenon is by no means to be interpreted by supposing that the water molecules require a minimum electromotive force of 1·7 volts to decompose them, but is due to polarization; a subject which cannot be treated here, and concerning which the student is referred to works on physical chemistry.

The Theory of Electrolytic Dissociation. The earlier theories of electrolysis, which attempted to explain the process as one consisting essentially of decomposition by electricity, followed by transference of the products to the electrodes, are inconsistent with the fact that electrolytic conduction takes place with the smallest difference of potential which can be applied to the electrodes, so long as the process is not vitiated by polarization. Clausius (1857) realized that such a decomposition would involve an expenditure of work, and to overcome this difficulty he assumed that a small fraction of the molecules of the solute are decomposed in solution into their constituents, and that these fragments convey the electricity through the solution to the respective electrodes. In other words, electrolysis consists of electric transference, and not of decomposition. Arrhenius (1887) correlated these views with the abnormal osmotic properties of electrolytes, and developed the theory of electrolytic dissociation or *ionization*. According to this theory, the molecules of an electrolyte such as sodium chloride are decomposed in solution to a considerable extent, and in dilute solution almost completely into sodium atoms, each with a high charge of positive electricity, and chlorine atoms correspondingly negatively charged. These charged atoms are known as *ions*; cations when positively charged and anions when negatively charged.

The conception of an ion therefore is that of an atom (or group of atoms acting as a whole, such as a radical) *plus* an electric charge, enormous in size relatively to the mass of the ion. These electric charges prevent the atoms or groups from manifesting their ordinary properties. For instance, the presence of chlorine ions does not impart to a solution any of the characteristics of chlorine gas, such as smell, colour, and bleaching properties. Again, the sulphate radical or (SO_4) group, although quite incapable of an independent existence *as such* when uncharged, does exist and is quite stable when accompanied by two units of negative charge*.

A solution of sodium chloride may be visualized as containing sodium chloride molecules, sodium ions, and an equal number of chlorine ions. The equilibrium between them can be written:

$$NaCl \rightleftharpoons \overset{+}{N}a + \bar{C}l,$$

indicating that there is a reversible chemical change between the molecules and the ions. As this change proceeds continuously, the equilibrium is dynamic and not static. If two strips of platinum each connected with one of the terminals of a battery be placed in such a solution, the ions follow the laws of attraction and repulsion. The negatively charged chlorine ions are drawn to the positively charged electrode, or anode, and on reaching it lose their charge to the electrode itself, becoming ordinary chlorine

---

* The number of units of electric charge carried by an ion is equal to its valency.

gas. In similar manner the sodium cations travel to the negatively charged electrode, or cathode, and are freed from their electric charge, becoming ordinary sodium. Since this metal at once reacts with water, it does not appear. What is actually observed is an evolution of hydrogen, while the liquid surrounding the cathode becomes alkaline from the sodium hydroxide formed. The cathodic products of the electrolysis of common salt are therefore hydrogen and caustic soda, due to secondary reactions. In very dilute solutions secondary reactions also take place at the anode. There is an interchange between the chlorine and the water, resulting in the liberation of oxygen.

When hydrochloric acid is electrolysed, chlorine and hydrogen are evolved in equal volumes, but since chlorine is somewhat soluble in hydrochloric acid, equal volumes of the gases cannot be collected until saturation with chlorine has been reached. Very dilute solutions, both of hydrochloric acid and of chlorides yield, not chlorine, but oxygen at the anode.

The so-called electrolysis of water is really the electrolysis of dilute solutions of acids, bases, and certain salts, since pure water is not an electrolyte, either or both of the gases being the products of secondary actions. Sulphuric acid ionizes thus in dilute solution:

$$H_2SO_4 \rightleftharpoons \overset{+}{H} + \overset{+}{H} + (\overset{=}{S}\overset{-}{O}_4),$$

the oxygen being the product of a secondary action between the discharged sulphate ion and the water:

$$(SO_4) + H_2O = H_2SO_4 + O.$$

From these equations it will be seen why two volumes of hydrogen and one volume of oxygen are obtained by the electrolysis of acidified water. It is also clear that the decomposition of water by an electric current is entirely due to secondary reactions.

In more concentrated solutions sulphuric acid ionizes into $\overset{+}{H}$ and $H\overset{-}{S}O_4$, and, as has been seen, under suitable conditions the $H\overset{-}{S}O_4$ ions can be released from their charge without decomposing, persulphuric acid, $H_2S_2O_8$, being formed in this way.

Copper sulphate, when electrolysed between platinum electrodes, yields metallic copper at the cathode and oxygen at the anode, the ions being $\overset{++}{Cu}$ and $(\overset{--}{SO_4})$. The solution surrounding the anode, in the absence of convection currents, becomes more and more acid as the electrolysis proceeds, and contains diminishing amounts of copper. If copper electrodes be used the concentration of the solution does not alter, and with a suitable current no gas is liberated. The $\overset{--}{SO_4}$ anion, on reaching the anode, removes an atom of copper from it, again forming copper sulphate. Thus the electrolysis results in a transference of copper from the anode to the cathode. This phenomenon is turned to account in the preparation of very pure 'electrotype' copper, and in the manufacture of weldless copper tubes. The anode in these cases is usually crude or 'blister' copper, containing impurities which ultimately find themselves at the bottom of the electrolytic cell as a muddy deposit, which is later worked up for silver and gold which nearly always accompany copper ores.

The truth of Faraday's law is demonstrated as a laboratory experiment by passing a current through three electrolytic cells arranged in series. The first contains acidified water, the second a solution of copper sulphate, and the third a solution of cuprous chloride in hydrochloric acid, platinum electrodes being used throughout. In the second and third cells the copper is deposited on the cathodes as a coherent film which, after washing and drying, can be weighed, the volumes of oxygen and hydrogen from the first cell being of course measured, and their weights calculated. The experiment is intended to shew that the masses of the liberated substances hydrogen, oxygen, copper (cupric), and copper (cuprous) are in the ratios 1 : 8 : 31·8 : 63·6, that is, they are proportional to the chemical equivalents of these elements, as the law requires.

The unit of electric quantity is the *coulomb*, or that quantity represented by a current of one ampere flowing for one second. When passed through an aqueous solution of silver nitrate, this quantity of electricity deposits 0·001118 grm. of silver. Since this particular measurement can be made with great accuracy, the definition is inverted, the coulomb being now defined as that quantity of electricity which deposits 0·001118 grm. of silver. This is the true electro-chemical equivalent of silver, and it follows from Faraday's law that the electro-chemical equivalent of any other ion (element or group) must be

$$= \frac{0 \cdot 001118 \times \text{equivalent weight}}{108},$$

108 being the equivalent weight of silver.

This may be stated in another way. In order to liberate one gramme of hydrogen in electrolysis, 96,540 coulombs are required; a quantity of electricity called one *faraday*. This is the quantity of electricity required to liberate one gramme-equivalent of *any* ion.

It has already been shewn that the osmotic pressure of a dilute solution of a non-electrolyte, such as cane sugar or urea, can be calculated from the concentration of the solution and the formula weight of the solute, and that the values so obtained are in close agreement with experimental determinations. In the case of electrolytes the observed values are always greater than those calculated.

Binary electrolytes (those yielding one cation and one anion only, such as sodium chloride) give values for osmotic pressure (or other effect directly proportional thereto, such as depression of the freezing point) approaching twice that calculated from the formula, when the solution is dilute. More concentrated solutions of common salt shew relatively smaller osmotic pressures, but these are still always greater than would be expected from a non-electrolyte. Osmotic effects are proportional, not to the number of molecules, but to the total number of independent particles (molecules *plus* ions) present in a given volume of the solution.

The difference between the calculated and the observed depression of the freezing point of a solution provides a means of determining the degree of ionization. If α be the fraction of the electrolyte which is ionized, then $1 - \alpha$ is the un-ionized fraction. When each dissociated molecule yields two ions, the total number of individual units (molecules *plus* ions) $= 1 - \alpha + 2\alpha$; or $1 + \alpha$.

Let p be the calculated osmotic pressure on the assumption that there is no dissociation, and p' the actual value observed. Then $p' = ip$, where i is a factor known as van 't Hoff's coefficient. Since the osmotic pressure is proportional to the total number of molecules *plus* ions present:

$$\frac{1 + \alpha}{ip} = \frac{1}{p}; \quad \frac{1 + \alpha}{i} = 1; \quad \text{or } \alpha = i - 1.$$

By making determinations of the value of α over a varied range of concentration, a dissociation-concentration curve may be obtained. This curve is asymptotic; that is, dissociation constantly increases with dilution without ever actually reaching completion, and this is what is meant by the statement that total ionization is only reached at infinite dilution.

In conductors of the second class, i.e. electrolytes, the electricity is carried solely by the ions, and therefore the conductivity is proportional to the number of ions present. Measurements of conductivity with the object of determining the degree of ionization in a solution are too complex for treatment here.

An acid is sometimes defined as a compound which yields hydrogen ions when dissolved in water. Acidity is in fact solely due to the presence of hydrogen ions, just as alkalinity is due to the presence of hydroxyl ions and nothing else. The neutralizing effect which acids and bases exert upon each other is nothing more than the union of the hydrogen ions with the hydroxyl ions to form water, the water molecules being almost undissociated. There is, indeed, a slight dissociation, estimated at about one water molecule per five hundred million, which is negligible for the present purpose though it has an importance which will be seen later.

Acids, like all other electrolytes, including bases, differ greatly in their degree of ionization. Those acids which are highly ionized are termed 'strong,' while those which are feebly ionized are described as 'weak.' *Strength* in this sense is dependent on the nature of the acid and must not be confused with *concentration*.

If dilute equivalent solutions of different acids be compared by any of the methods mentioned above for determining the degree of ionization, it will be found possible to arrange them in the order of their strength. The order in which they appear is the same over a considerable range of dilution, though all acids must have the same strength at infinite dilution. In this sense nitric and hydrochloric are the strongest acids and are approximately equal, sulphuric acid having about half the strength of these, while acetic acid is very much weaker. The weakest of all are boric, hydrocyanic, and carbonic acids. In the same sense potassium and sodium hydroxides are the strongest bases, while a solution of ammonia in water is but a feeble base.

The concentration of a solution can be expressed in several ways. In the case of the dilute solutions required in volumetric analysis, the concentration is best expressed as mass of solute in unit volume of solution. Any solution whose concentration is accurately known is a *standard* solution. A *normal* solution is defined as one which contains one gramme-equivalent of the solute per litre of solution. For instance, a normal solution of sulphuric acid contains 49 grm. of H_2SO_4 per litre, since 49 grm. of this acid contains one gramme of acid hydrogen. Similarly, normal solutions of potash and soda contain respectively 56 grm. of KOH and 40 grm. of NaOH per litre.

Theoretically, a given volume of a normal solution of any acid should always neutralize an equal volume of any normal alkali, notwithstanding the different 'strengths' of acids. From what has been stated above it will readily be understood that, say, 100 c.c. of normal acetic acid, a weak acid, contain fewer hydrogen ions than the number of hydroxyl ions contained in 100 c.c. of normal potash, a strong base, and that mixing these two equal volumes would produce a solution containing an excess of hydroxyl ions, and therefore having an alkaline reaction. Such is not the case however. A distinction must be drawn between *actual* and *potential* ions. When the actual ions of the acetic acid are converted into water by union with the hydroxyl ions of the base, more of the undissociated acetic acid molecules ionize in an attempt to restore equilibrium, and this proceeds until *either* there are no more hydroxyl ions left *or* the whole of the acetic acid molecules have ionized. Both these conditions are fulfilled when equivalent masses of potash and acetic acid are present, and a neutral solution results.

Indicators. An indicator is a solution which changes colour according as it is made acid or alkaline. Practically all the indicators in common use are *acid* indicators, so called because they are themselves feeble acids, which hardly ionize at all. Their metallic salts ionize considerably however. All such indicators possess molecules differently coloured from their anions. The addition of a slight excess of soda to a solution containing an indicator results in the formation of the sodium salt, which on ionizing yields an anion whose colour is readily distinguished from that of the undissociated molecule of the indicator itself. As an example, phenolphthalein is a weak acid which may be represented as H.Phth. The undissociated molecule is colourless, but in presence of soda the readily ionized sodium salt is formed, Na.Phth. This yields an anion, Phth, of a deep crimson colour. On the addition of an acid this colour is of course again discharged, in consequence of the re-formation of the undissociated acid molecule H.Phth.

Hydrogen Ion Concentration. Hydrogen ions play a most important part in biological and biochemical problems. Micro-organisms in general can only thrive when the concentration of hydrogen ions in the medium in which they are found lies between certain limits, which vary with the organism. Pure water is a neutral liquid, but as has been mentioned, it ionizes to a slight extent. *True neutrality* does not lie in the *absence* of hydrogen or hydroxyl ions but in the equality of their concentration. At 18° C. the concentration of hydrogen ions and also of hydroxyl ions in water is about $10^{-7\cdot07}$ gramme-ions per litre, though as the temperature rises this concentration increases considerably. Pure water may therefore be regarded as both an acid and an alkali, of about one ten-millionth normal concentration; that is, it contains one gramme of hydrogen ions and 17 grm. of hydroxyl ions in ten million litres. If the concentration in gramme-ions per litre be represented by C_H and C_{OH} respectively, the equilibrium between the ions and the undissociated water molecules, $C_{(H_2O)}$, is, according to the law of mass action, expressed by the equation:

$$C_H \times C_{OH} = K (C_{H_2O}).$$

Since the concentration of un-ionized water is so enormous as compared with that of the ions it may be assumed to be invariable, and the equation may be written:

$$C_H \times C_{OH} = K,$$

where K, as determined by different experimental methods, has a value of $10^{-14 \cdot 14}$ at 18° C.

For convenience hydrogen ion concentration is represented by the symbol P_H, which denotes the logarithm of C_H with the negative sign omitted. Thus for pure water $C_H = 10^{-7 \cdot 07}$, so $P_H = 7 \cdot 07$. As the sum of the indices representing the hydrogen and the hydroxyl ion concentration must always be $- 14 \cdot 14$, a solution having $C_H = 10^{-8}$ must have $C_{OH} = 10^{-6 \cdot 14}$. That is, if $P_H = 8$, $P_{OH} = 6 \cdot 14$.

In decinormal acetic acid solution the degree of ionization of the acid is found experimentally to be 1·36 per cent. Since decinormal acid solutions all contain 0·1 grm. of acid hydrogen per litre, and since in this case 1·36 per cent. of it is in the ionic state, the concentration of hydrogen ions, C_H, is 1·36 per cent. of 0·1 grm.

= 0·00136 grm.   = 1·36 × 10^{-3} grm. per litre.

As 1·36 = $10^{0 \cdot 134}$, 1·36 × 10^{-3} = $10^{0 \cdot 134}$ × 10^{-3}, = $10^{-2 \cdot 866}$ and therefore P_H for decinormal acetic acid = 2·87 (nearly). Again, in decinormal ammonia solution it can be shewn experimentally that the compound NH_4OH ionizes to the extent of 1·3 per cent. That is, 1·3 per cent. of one-tenth of a gramme-equivalent of (OH) ions are present in one litre of solution, and therefore

$$C_{(OH)} = \frac{1·3}{10 \times 100} = 0·0013.$$

But $0·0013 = 1·3 \times 10^{-3}$; and since $1·3 = 10^{0·11}$

$$1·3 \times 10^{-3} = 10^{0·11} \times 10^{-3} = 10^{2·89}.$$

Therefore for $\frac{N}{10}$ ammonia solution, $P_{(OH)} = 2·89$, and

$$P_H = 14·14 - 2·89, = 11·25.$$

It follows that values for P_H numerically less than 7·07 indicate what is usually meant by acidity, while values greater than 7·07 indicate alkalinity. The following table gives the P_H values for some typical liquids.

Liquid	Value of P_H
Decinormal hydrochloric acid	1·04
Decinormal acetic acid	2·87
Milk	6·9
Water	7·07
Blood	7·35
Decinormal ammonia	11·27
Decinormal soda	13·1

Indicators in general do not by any means necessarily register true neutrality. It has been seen, for example, that phenolphthalein does not change colour until the concentration of hydroxyl ions slightly exceeds that of the hydrogen ions. Different indicators shew their particular colour change at different P_H values. In recent years much attention has been given to the preparation of indicators whose maximum colour change corresponds with some particular P_H value, and a series of such is now available which covers almost the whole range of the table given above.

Hydrolysis. Sodium carbonate has been previously described as a normal salt with an alkaline reaction. Since the latter implies the presence of $\overset{-}{O}H$ ions the salt must have reacted with water to produce them: in other words it has hydrolysed.

Sodium carbonate in solution ionizes thus:

$$Na_2CO_3 \rightleftharpoons \overset{+}{N}a + \overset{+}{N}a + (\overset{--}{C}O_3).$$

Water itself produces a minute concentration of $\overset{+}{H}$ and $\overset{-}{O}H$ ions. Since carbonic acid itself is almost undissociated the $\overset{+}{H}$ ions unite with the $(\overset{--}{C}O_3)$ ions to form carbonic acid, which leads to a further ionization of water in an attempt to restore the equilibrium. As the $\overset{-}{O}H$ ions are not so removed, sodium hydroxide being a highly ionized compound, it follows that there is a preponderance of $\overset{-}{O}H$ ions in the solution when equilibrium is reached, hence the alkaline reaction.

Potassium cyanide hydrolyses similarly with water. The solution has an alkaline reaction and smells of prussic acid, HCN. The salt ionizes into $\overset{+}{K}$ and $(\overset{-}{C}N)$ and these react with the $\overset{+}{H}$ and $(\overset{-}{O}H)$ ions from the water, producing the undissociated acid and highly ionized soda.

In these two instances it will be noticed that the salts are derived from strong bases and weak acids. This kind of hydrolysis is typical of all salts so derived. On the other hand, salts produced from weak bases and strong acids, such as copper sulphate and ferric chloride, when dissolved in water shew an acid reaction. Ferric chloride, for instance, dissociates into $\overset{+++}{Fe}$ and $3\overset{-}{C}l$. Since hydrochloric acid is highly ionized there is but little tendency towards the union of the $\overset{+}{H}$ of water with the $\overset{-}{Cl}$, while the union of the $\overset{+++}{Fe}$ with the $(\overset{-}{OH})$ of water proceeds to a considerable extent, ferric hydroxide being practically undissociated. This implies a preponderance of hydrogen ions in the solution.

The acidity of a solution of ferric chloride cannot be neutralized by an alkali. The addition of hydroxyl ions would lead to a precipitation of ferric hydroxide, but so long as any ferric chloride remained in solution there would be an acid reaction. Alkalinity of the solution could only coincide with the total removal of the iron by precipitation.

If two salts such as potassium nitrate and sodium chloride be dissolved together in water, there will be present four ions, viz. $\overset{+}{Na}$, $\overset{+}{K}$, $\overset{-}{NO_3}$ and $\overset{-}{Cl}$. This would entail the presence of four different kinds of molecules, KNO_3, $NaNO_3$, KCl, and $NaCl$, all eight individuals coexisting in dynamic equilibrium. Solutions identical in every respect would be obtained by taking equivalent quantities *either* of sodium chloride and potassium nitrate, *or* of potassium chloride and sodium nitrate.

The usual processes of precipitation, such as take place in qualitative analysis for instance, turn upon the interaction of ions. It may be said in general that if a solution contain ions which by their union can produce an insoluble (and undissociated) compound, this is what will take place, and will continue so long as both kinds of ion are present. For instance, if a solution of silver nitrate be added to a solution of common salt, the four ions present are $\overset{+}{Ag}$, $\overset{+}{Na}$, $\overset{-}{NO_3}$, and $\overset{-}{Cl}$. When $\overset{+}{Ag}$ meets with $\overset{-}{Cl}$, insoluble (and undissociated) silver chloride, $AgCl$, is produced and is removed from the sphere of action as a white precipitate, taking no further part in the change. Excess of silver nitrate removes the whole of the chlorine ions, actual and potential, and in the same way excess of sodium chloride removes *all* the silver from solution.

It will now be understood that soluble compounds containing chlorine do not of necessity give a precipitate with silver nitrate solution. Chloroform and potassium chlorate, for instance, give no such precipitate. The chlorine must be present in the $\overline{Cl}$ condition. The $\overline{ClO_3}$ anion from potassium chlorate, $KClO_3$, can, of course, unite with the silver ion, but it then produces soluble silver chlorate, $AgClO_3$, and not silver chloride.

On the basis of the ionic theory it is not the most insoluble compound which is formed in these cases, but the one least dissociated. In most instances the least dissociated happens to be the most insoluble compound but not always, as otherwise it would in general be impossible to dissolve a precipitate. The addition of sodium sulphide to a solution of zinc chloride results in the precipitation of insoluble zinc sulphide, because zinc sulphide is not only the most insoluble combination of ions possible in this case, but it is also the least dissociated. Zinc sulphide, however, dissolves in hydrochloric acid, and a consideration of the equation, $ZnS + 2HCl = ZnCl_2 + H_2S$, will shew that zinc sulphide is still the most insoluble combination of the ions present in the system. The direction of the reaction seems to be determined by the very small dissociation of hydrogen sulphide.

Zinc sulphide is readily soluble in hydrochloric acid but not in acetic acid. This was formerly explained qualitatively by saying that the 'strong' hydrochloric acid can dissolve the solid, but that the 'weak' acetic acid cannot do so. The conception of *solubility product* is helpful in this connection.

If a very sparingly soluble salt be placed in water (no

salt is absolutely insoluble) the equilibrium between the ions and the undissociated molecules is regulated by the law of mass action, which for the present purpose may be expressed $ab = kc$, where a represents the concentration of the cations, b that of the anions, c that of the undissociated molecules, k being a constant. If either a or b is increased the product kc is likewise increased, and as the solution is saturated to begin with, any increase in kc will result in the precipitation of some of the compound from solution (cf. p. 265).

Now if zinc sulphide is placed in water, there is an equilibrium between zinc ions, sulphide ions, and undissociated zinc sulphide. Addition of hydrogen ions from a strong acid such as hydrochloric will result in the formation of undissociated hydrogen sulphide, followed by further ionization of zinc sulphide. As the sulphide ions are removed in this way, more zinc sulphide must pass into solution to keep up the supply of sulphide ions, or in other words, the compound dissolves. With acetic acid the concentration of hydrogen ions is so small that the formation of undissociated hydrogen sulphide scarcely takes place, so this acid does not dissolve the salt.

If a solution of ammonia be added to a solution of magnesium sulphate an insoluble hydroxide of magnesium is produced. This precipitation takes place in consequence of the solubility product of magnesium hydroxide being overstepped, the hydroxyl ions being supplied by the ammonium hydroxide:

$$MgSO_4 + 2NH_4OH = Mg(OH)_2 + (NH_4)_2SO_4.$$

If, however, ammonium chloride be added to the solution of the magnesium salt, subsequent addition of ammonia does not produce a precipitate. The effect of adding ammonium chloride is therefore to diminish the precipitating power of the base, and this is considered to be due to the diminution

of the concentration of the hydroxyl ions. Ammonium
chloride being a good electrolyte is highly ionized,

$$NH_4Cl \rightleftarrows (N\overset{+}{H}_4) + \overset{-}{Cl},$$

while the hydroxide NH_4OH is a weak base and only
slightly dissociated:

$$NH_4OH \rightleftarrows (N\overset{+}{H}_4) + (O\overset{-}{H}).$$

There is therefore a large preponderance of ammonium ions,
and consequently a diminution in the concentration of
hydroxyl ions, which thus falls below the concentration
necessary for precipitation of magnesium hydroxide.

This suppression of ionization, shewn by the restraining
effect of ammonium chloride or other similar ammonium
salt on the dissociation of ammonium hydroxide, is important in analytical operations. Briefly, the addition of
ammonium chloride is equivalent to increasing the concentration of ammonium ions without introducing other
ions which might complicate the reaction, and this results
in the suppression of the dissociation of the ammonium
hydroxide, thus diminishing the concentration of hydroxyl
ions. This diminution of ionization can be made visible in
certain cases where coloured ions are concerned. Cupric
ions are blue, and therefore all cupric salts which have
colourless anions, such as the chloride, nitrate, sulphate,
etc., will give blue solutions. This colour may be modified
by that of the undissociated molecule, but all will shew the
same blue colour when sufficiently dilute. The cupric
chloride *molecule* is brown, and in fairly concentrated
solutions this colour modifies that of the cupric ions, producing an intermediate green. This, on dilution, passes
through a greenish blue to pure blue. If concentrated
hydrochloric acid be added to a solution of cupric chloride,
the ionization of the latter is suppressed by the chlorine

ions from the acid. More undissociated molecules of cupric
chloride are produced at the expense of the cupric ions, and
the change in colour is very noticeable.

The hydrogen ion concentration of a strong acid, such as
hydrochloric, can be greatly diminished by the addition of
the sodium salt of a weak acid, such as sodium acetate.
A study of the equation:

$$\overset{+}{H} + \overset{-}{Cl} + \overset{+}{Na} + CH_3\overset{-}{C}OO = CH_3COOH + \overset{+}{Na} + \overset{-}{Cl},$$

will shew that the tendency is towards the formation of
acetic acid, since this is the least dissociated of the possible
compounds, with the removal of hydrogen ions. Here again,
careful distinction must be drawn between actual and
potential ions. While the acidity of the solution in the
sense of hydrogen ion concentration is greatly diminished
by the addition of sodium acetate, the mass of sodium
hydroxide which would be required for neutralization is in
no wise modified by the addition of sodium acetate.

From the standpoint of the ionic theory, processes of
oxidation, such as the conversion of ferrous salts into ferric,
are regarded as due to increase of the cationic charge. The
converse process, reduction, is similarly considered as a
decrease of charge on the positive ion. Thus, ferric ions
are written $\overset{+++}{Fe}$ and ferrous $\overset{++}{Fe}$. The mutual oxidation and
reduction which takes place between solutions of ferric and
stannous chlorides, usually represented thus:

$$2FeCl_3 + SnCl_2 = 2FeCl_2 + SnCl_4,$$

may be written as an ionic equation:

$$2\overset{+++}{Fe} + \overset{++}{Sn} = 2\overset{++}{Fe} + \overset{++++}{Sn}.$$

As already explained, complex salts are quite distinct
from double salts. Solutions of the latter are identical with

solutions prepared by mixing equivalent quantities of the simple salts; that is, the ions of the double salt are those of the simple salts from which the double salt is derived. The complex salt potassium ferrocyanide, $K_4FeC_6N_6$, yields four potassium ions and a quadrivalent ferrocyanide anion, $\overline{Fe(CN)}_6$. When potassium ferrocyanide is oxidized by chlorine, potassium chloride and potassium ferricyanide, $K_3FeC_6N_6$, are produced:

Molecular equation,

$$Cl_2 + 2K_4FeC_6N_6 = 2KCl + 2K_3FeC_6N_6.$$

Ionic equation,

$$Cl_2 + 2\overline{Fe(CN)}_6 = \overline{Cl} + \overline{Cl} + 2\overline{Fe(CN)}_6.$$

The oxidation consists in the removal of one electro-negative charge from the quadrivalent ferrocyanide ion, with its consequent transformation into the tervalent ferricyanide ion, $\overline{Fe(CN)}_6$, the free chlorine being converted into the ionic condition. From this point of view, the oxidation of a *cation* lies in *increasing* the electro-positive charge upon it, while the oxidation of an *anion* consists in *decreasing* its electro-negative charge.

The ferrocyanide and ferricyanide ions shew much greater stability than do complex ions in general. Iodine dissolves freely in solutions of potassium iodide with the formation of univalent complex iodine anions, $\overline{I}_3$, which have a brown colour. These are very easily decomposed into simple univalent colourless iodine anions, $\overline{I}$, and molecules of free iodine, I_2.

The ionic theory gives a clear and intelligible picture of many of the phenomena of solution which without it would be difficult to understand. It is now very generally accepted as a basis on which reactions in solution may be explained, but it should not be forgotten that it *is* a hypothesis and nothing more, and is in some respects defective. There are many reactions known for which the ionic theory offers no satisfactory explanation, and some to which it seems directly opposed, but in spite of this no other theory of solution has approached it in general usefulness.

Chapter XI

METALS AND THEIR COMPOUNDS

The division of the elements into metals and non-metals, always an arbitrary and unsatisfactory classification, is really of minor importance. The metallurgist and the chemist view metals from somewhat different angles. The former lays stress upon certain physical properties, while the latter is more particularly concerned with the chemical reactions of the element and its compounds, and especially with the properties of the oxides. Metallic characteristics, although difficult to define, are nevertheless in many cases very pronounced. The general idea of this classification turns upon the sum of such properties as lustre, malleability, and ductility, and the power of forming mixtures with each other known as alloys. Such elements as are undeniably metallic are all very good conductors of heat and electricity, and they each form at least one basic oxide. As a class, metals unite with hydrogen much less readily than the non-metals, while the compounds thus formed, the hydrides, are more easily produced and are much more stable among the non-metals than among the metals.

Metals are generally found naturally in combination, though a few occur in the free condition. Gold and platinum almost always do so, either as nuggets or 'dust' in alluvial deposits, or disseminated in microscopic or ultra-microscopic particles in quartz rocks. A compound of gold with tellurium found in Australia is little more than a curiosity. On the south shores of Lake Superior large masses of uncombined copper have been found and worked.

The most abundant ores of heavy metals are the sulphides and oxides; those of the alkaline earths, magnesium, calcium, strontium, and barium, are the carbonates and sulphates, while the alkali metals sodium and potassium occur most freely as chlorides.

The term 'ore' as used by metallurgists signifies a compound from which the metal can be economically extracted. Thus, while sand usually contains iron it is not considered as an ore of iron, since the metal cannot be recovered profitably. A certain amount of earthy matter usually accompanies an ore as it is mined and this is known as the gangue. In many smelting operations the gangue has to be converted into a substance which will melt and flow before it can be removed. To this end a flux is added; that is, something which will react with the gangue to produce a fusible mass known as a slag. Gangue + flux = slag.

The extraction of metals from their ores is usually (though not always) carried out at high temperatures, which are generally well above the fusing point of the metal itself. For this purpose furnaces of various types are used, each metal and even each process having in course of time developed from experience an adaptation of the furnace peculiar to itself. The various types may be classified as follows:

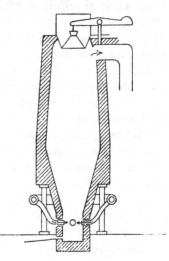

Fig. 37

1. Those in which the ore, the flux, and the fuel are all mixed together. A good example of this is the blast furnace used in iron smelting (Fig. 37).

2. Those in which the material to be heated comes in contact with the products of combustion but not with the solid fuel. These are reverberatory furnaces, the heat being reflected downwards from above on to the hearth, which is either fixed (Fig. 38), or of the tubular shape which can be revolved by machinery.

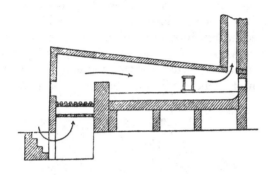

Fig. 38

3. Those in which a high temperature can be attained but where the material to be heated is not in contact with either fuel or the products of combustion. Such are crucible, retort, or muffle furnaces (Fig. 39).

4. Electro-thermal furnaces, used when the highest temperatures are required, the source of heat being the electric arc, usually between carbon poles (Fig. 40).

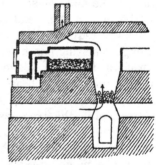

Fig. 39

The actual winning of the metal can seldom be accomplished in one operation; many processes are generally involved, such as calcining, roasting, smelting, and refining. There is very little difference between roasting and calcining. The material is heated to a moderate temperature, usually with access of air. The object of this operation is to remove impurities which can be rendered volatile, such as water, sulphur, and arsenic, and often to convert the metal sought into its oxide. When low temperatures are sufficient, revolving furnaces are used. For high temperatures, fixed hearth furnaces are more convenient.

The process of smelting generally consists in heating a mixture of the calcined ore with a reducing agent, usually some form of carbon, and when necessary a flux. Reduc-

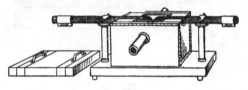

Fig. 40

tion to the metallic state takes place, and the metal which settles in a molten condition below the slag is run off, but generally requires a further process of purification or refining. Smelting can be carried out in any of the types of furnaces described above. Retort furnaces are used for volatile metals which are distilled, such as cadmium and zinc, blast furnaces for iron, and sometimes for lead and copper, while the fixed hearth is generally employed in the case of tin, lead, and copper.

In reduction methods the reducing agent is either carbon or carbon monoxide, the ore having been converted into oxide, wholly or in part where necessary, by a previous calcination. The chemistry of the blast furnace process for obtaining iron consists essentially in the conversion of coke into carbon monoxide which then reduces the ferric oxide to the metal. Solid carbon no doubt takes part in the reduction. The crude product, known as pig iron, contains as impurities considerable quantities of carbon, partly in combination as carbide of iron and partly as graphite. Silicon is also present and frequently other metals. The purification of pig iron is a process of removing carbon and silicon by oxidation, either by blowing hot air through the molten metal when the impurities are the first to be attacked, or by heating to fusion in presence of haematite, Fe_2O_3, or magnetic oxide, Fe_3O_4, whereby the carbon and silicon are removed, the former as gas, the latter in the slag. Steel in its various forms is iron containing carbide of iron, ranging from 0·1 per cent. of carbon in the case of mild steel up to 1·5 per cent. or more in the case of high carbon steels. Many of the special steels now in use, some of them very valuable on account of their extreme hardness at high temperatures, are alloys of steel with other metals, such as nickel, manganese, chromium, tantalum, tungsten, molybdenum, and vanadium.

The ore worked for tin is practically always the oxide, SnO_2. In this case the reduction proceeds very easily and at a comparatively low temperature. On heating a mixture of tinstone and fine coal the metal separates. This is purified by stirring the molten mass with a pole of wood, the gases from which reduce traces of oxide. The reduction of zinc ore differs from that of tin only in so far as the zinc distils away from the mixture of ore and carbon, to be subsequently condensed. The winning of mercury from the sulphide ore, cinnabar, HgS, is accomplished by simple roasting. The sulphur is oxidized to the dioxide, the metal volatilizing into condensers. Sulphide ores are generally subjected to a roasting process, which converts them to the oxide, before being smelted with a reducing agent, though frequently this preliminary oxidation is not carried to completion. In the case of copper sulphide ores the percentage of the metal is generally not very high, but by calcination much of the iron present may be converted into the oxide, while the copper remains as cuprous sulphide, Cu_2S. Smelting with a suitable flux removes the iron as fusible silicate, while the cuprous sulphide sinks beneath in a molten condition, technically known as 'fine metal.' This is later ground to small pieces and roasted to the oxide, CuO. On mixing this oxide with the calculated weight of cuprous sulphide and heating, the following reaction takes place:

$$Cu_2S + 2CuO = 4Cu + SO_2,$$

the copper being reduced to the metallic condition. As this is run off the escaping sulphur dioxide throws the solidifying metal into blisters—hence the name, blister copper.

Blister copper is very impure, and is used either for making blue vitriol, bluestone, or copper sulphate; or as the anode in producing electrolytic copper; or it is purified by poling. This consists in introducing a green sapling into the molten metal, when the gases from the charring wood reduce the dissolved oxide to the metal. The agitation in the metal during this process is violent, and no doubt assists in the separation of entangled slag.

Blue vitriol is made from blister copper on a large scale for agricultural use and for other purposes. When the metal is required for the sulphate, it is run while still molten into cold water and so granulated, the granules taking the shape of thin hollow spheres which vary in size from that of peas to that of walnuts. These are subjected to the action of steam and air for a few minutes, whereby they become coated with a thin film of oxide. They are then lowered into hot dilute sulphuric acid when the oxide is rapidly dissolved off. This alternate process of oxidation, with steam and air and dissolving away the oxide produced, is continued until the little copper spheres disintegrate. The mud which settles in the sulphuric acid tanks is worked for the silver and gold which practically always accompany copper ores. A similar mud is obtained when blister copper anodes are used in the electrolytic process.

The oxidation of a metallic sulphide sometimes, though by no means always, produces a sulphate. Iron pyrites, FeS_2, is not usually worked for iron directly, but ferrous sulphate is obtained from it. Large quantities of very impure pyrites are brought up from certain coal mines, piled up in heaps together with scrap iron, exposed to the action of the weather and occasionally watered. Some free sulphuric acid is formed, which dissolves the scrap iron, but the whole reaction may be summarized thus:

$$FeS_2 + Fe + 14H_2O + 4O_2 = 2FeSO_4 . 7H_2O.$$

The liquid which drains away from the heaps is allowed to settle, the clear solution drawn off and evaporated, and the ferrous sulphate crystallized out. This compound is known technically as copperas, or green vitriol.

Dry lead sulphide is oxidized to sulphate at higher temperatures, and by careful regulation of the air supply and the temperature the process can be made almost quantitative. For this purpose finely divided galena, natural lead sulphide, PbS, is allowed to fall down a tower up which a stream of hot air is driven. The product is used for making paint.

The extraction of lead from its sulphide ores follows the same direction to some extent. The roasted ore becomes lead sulphate in part, though most of it is converted into the oxide, the roasting being continued until an excess of oxide is formed. On raising the temperature of the hearth and cutting off the air supply the following interactions take place:

$$PbS + PbSO_4 = 2Pb + 2SO_2,$$

$$PbS + 2PbO = 3Pb + SO_2.$$

The addition of a little fine coal results in the reduction of the excess of lead oxide.

Poor lead ores are usually roasted until the metal is completely converted into the oxide. This is then mixed with coke and iron oxide ore and smelted in a blast furnace, the lead being reduced to the metallic state and the siliceous gangue converted into fusible ferrous and calcium silicates.

Another method of lead extraction turns upon the re-placement of the lead from the sulphide by the more electro-positive metal iron. The ore is heated with a flux which converts the gangue into a fusible slag, while the heavier galena fuses and sinks below it. This is then treated with scrap iron which liberates the lead, the easily fusible ferrous sulphide formed passing into the slag. Instead of scrap iron, iron ore and coke are sometimes used.

A very similar process is almost exclusively used for antimony. The native sulphide, stibnite, Sb_2S_3, is melted out of the mother rock. Scrap iron is introduced into this, when the liberated antimony sinks below the fused ferrous sulphide and can then be run off.

The ingots of lead from smelting operations are gener-ally 'hard' and are always impure. Furthermore, few specimens of lead ores are free from silver, and most of them contain sufficient of this metal to make the extrac-tion a profitable process. The usual impurities in lead are metals, the commonest being antimony. Purification, or softening as it is called, is effected by oxidation. Unlike copper, which dissolves some cuprous oxide, thereby becoming somewhat brittle, lead does not dissolve its oxide. On submitting the molten metal to a current of air most of the injurious impurities are the first to be attacked, but some lead is also oxidized. These oxides float as a scum on the surface of the molten metal and can be easily removed. The softening is regarded as complete when a specimen of the lead presents a flaky appearance. The scum is afterwards worked up and used for the preparation of typemetal. Since lead is as readily oxidized as antimony, it seems doubtful whether the explanation given above can apply to this metal. Antimony undoubtedly is removed, practically completely, but the nature of its combination seems uncertain.

The desilverizing of lead is carried out by two main processes, the older of which is due to Pattinson. As has been seen (p. 253), when a dilute solution is cooled the pure solvent is the first to crystallize, and by removing this the remaining solution can be made more concentrated. This is the principle involved, the lead being the solvent and the silver the solute. The plant consists of ten or more large pots arranged in a row, in which the lead can be melted and allowed to cool. A preliminary assay of the silver content of the lead determines into which of the pots it shall be placed. As the metal cools it is stirred with a perforated iron ladle, and the solidifying portions withdrawn and placed in the next pot. Theoretically this should be pure lead, but so much of the solidifying metal is entangled that only a partial separation is obtained. Two-thirds of the contents of the pot are so removed and placed in the adjacent pot to the right. The remaining one-third containing the solute (silver) is removed while still molten into the adjacent pot to the left. This operation is repeated on each pot as soon as it becomes full. Thus there is a stream

of a gradually concentrating solution of silver in lead proceeding leftwards, while purer and purer lead proceeds to
the right. After the last fusion the lead contains a quarter
of an ounce of silver or less per ton, and is known as D-lead.
No further purification is needed, as all other impurities
are removed with the silver. In fact, D-lead is said to be
one of the purest metals on the market.

The concentration of the silver is not usually carried
much beyond 300 ounces per ton by this process. At this
stage it is more economical to remove the remaining lead
as oxide. This is done by cupellation, i.e. the molten lead is
heated to redness in a strong current of air upon a cupel
made of bone ash or magnesia. The oxide forms an oily
looking liquid which is blown over the lip of the cupel into
receiving vessels. This is the source of practically all the
litharge, PbO, in the market. The process is stopped when
the lead has been reduced to one-tenth of its bulk, that is,
when it contains about 3000 ounces of silver per ton. The
final process does not differ from the above, but a new
cupel and a different furnace are generally used, because
the litharge formed at this stage contains enough silver to
be worth recovering. The disappearance of the last trace
of oxide of lead from the surface of the molten silver is
a very striking phenomenon known as the 'flashing' of
silver.

A more modern desilverizing method, known as Parkes's
process, turns upon the principle of the partition coefficient
(p. 107). The two non-miscible liquids in this case are the
molten metals lead and zinc, the solute, silver, being much
more soluble in zinc than in lead. Several tons of the crude
lead are melted in a crucible and zinc is added. The molten
mass is well stirred; a somewhat difficult process on account
of the difference of specific gravity of the two metals. On
standing, the zinc floats to the surface and solidifies be-

fore the lead. The solid is skimmed off, bringing much lead with it. The lead is twice treated with zinc, the first treatment being always with a charge of zinc from a previous operation, and the second with new zinc. The desilverized lead from this process requires softening as above described, since zinc dissolves to a slight extent in lead. The mass of zinc, with the entangled lead, is heated upon a sloping hearth at a temperature between the melting points of lead and zinc. This liquation allows much of the lead to melt and flow away, to be returned to the desilverizing pot. The residue is heated in a retort whereby the greater part of the zinc is recovered by distillation, the remainder being cupelled as before.

The isolation of a metal from one of its compounds by replacement by another metal has in the past received a wider application than at present. The earlier preparation of aluminium was effected by the action of sodium on the fused chloride. Moissan obtained calcium and other alkaline earth metals in an analogous manner, by fusing calcium iodide with sodium. During last century small quantities of very pure copper were obtained from the 'blue water' or dilute copper sulphate found in the island of Anglesea, by displacement of the copper with scrap iron. This was known as Mona copper, from the ancient name of the island.

In 1898 Goldschmidt found that an intimate mixture of ferric oxide and aluminium powder would ignite and react with great violence, attaining a temperature estimated to approach that of the electric arc. This mixture is known as thermite, and was used between 1914 and 1918 as a filling for incendiary shells. The energy of oxidation of aluminium is much greater than that of other metals, so that, once it is started, the reaction,

$$M_xO_3 + 2Al = Al_2O_3 + xM,$$

is strongly exothermic. The method has now become a general one for preparing metals which could not otherwise be readily obtained from their oxides or sulphides, such as chromium, manganese, and uranium. It is also used for attaining high temperatures, such as are required for welding steel rails. The reaction may be illustrated by placing an intimate mixture of ferric oxide and aluminium powder, in the calculated proportions, in a fireclay crucible. A little barium peroxide is placed in a small cavity on the surface of the mixture and a strip of magnesium ribbon introduced as a fuse. The burning magnesium in contact with the barium peroxide starts the reaction, and soon the whole contents of the crucible attain a white heat. The iron remains as a fused ingot.

The beginning of the electrolytic isolation of metals dates from the experiments of Davy, who obtained the metals sodium and potassium by electrolysis of their fused hydroxides in 1807. The method used by Davy is practically that which now obtains. The electrolysis of the fused *chlorides* of sodium and potassium has been tried, but has been found less satisfactory. The hydroxide is fused in an iron crucible which itself forms the anode, the cathode being a rod of iron.

When the chloride of a metal is readily fusible it can be conveniently used, but generally some slight modification is made. In the case of magnesium, for instance, the double chloride of magnesium and potassium (carnallite) is employed, since it occurs naturally in abundance and is not inferior to the pure chloride for the preparation of the metal. When chlorides are electrolysed the anode is made of carbon, the cathode usually being iron.

Aluminium is manufactured on a very large scale, particularly at places where hydro-electric power is available, as at Kinlochleven. The electrolyte (bauxite,

Al_2O_3) is dissolved in a bath of fused cryolite, Na_3AlF_6. The process is both electrolytic and electrothermic in the sense that there is a high resistance, and much current is expended in maintaining the temperature necessary to keep the charge in a molten condition. Bauxite is added from time to time, the solvent cryolite not being itself appreciably electrolysed. The oxygen liberated at the carbon anodes rapidly burns them to carbon monoxide and carbon dioxide.

Many experiments have already been described in which a metal has been displaced from one of its compounds by another metal. An interesting series of such experiments can be devised by taking solutions of various metallic salts and trying the effect upon them of different metals. For instance, if strips of zinc or iron or magnesium be immersed in dilute copper sulphate solution, the copper is obviously turned out of combination, while the metal used gradually disappears. On the other hand, a strip of copper immersed in a solution of silver nitrate or mercurous nitrate becomes coated with silver or mercury, some of the copper going into solution, as may be seen by the development of a blue colour. If a piece of zinc be covered with excess of lead acetate solution, a 'tree' of displaced lead appears, and ultimately the whole of the zinc is dissolved.

The experiments suggested above would indicate that the metals might be arranged in order of their reactivity in this sense, i.e. any metal in the column would displace any of those following it from solutions of their salts, while it, in turn, could be displaced from solutions of its salts by any metal preceding it in the column. Such a table is given on p. 289.

The mechanism of these changes may be represented in accordance with the ionic theory, thus:

$$Zn + \overset{++}{Cu} = Cu + \overset{++}{Zn}.$$

The tendency of one metal to displace another, that is, to take over its ionic charge, leaving the displaced metal uncharged, is due to differences in the electro-chemical character of the metals themselves.

The displacement of hydrogen from dilute sulphuric acid by zinc is another example of the same phenomenon, essentially electrolytic, as may be shewn thus. If a strip of *pure* zinc be placed in *pure* dilute sulphuric acid no appreciable action takes place. If a strip of platinum be also introduced into the acid so as not to touch the zinc, still no action takes place. If the two metals be then joined externally by a wire, hydrogen is at once evolved at the surface of the *platinum*, no gas being observed to come from the zinc. The zinc, however, passes into solution and a current of electricity flows through the wire. Plainly the hydrogen ions of the acid are attracted to the platinum strip and are there relieved of their charge, with liberation of hydrogen gas. The $\overline{\overline{SO}}_4$ ions are similarly attracted to the zinc strip, where they may be regarded as yielding up their charge and at the same time combining with some of the zinc to form zinc sulphate.

Such a cell is of little practical use however, since the current falls off very rapidly owing to polarization. The hydrogen bubbles attach themselves to the platinum electrode and impede the current, both by covering the platinum surface with a badly conducting film of gas, and by setting up an opposing electromotive force.

From the above experiment it appears that zinc has a stronger tendency to assume the ionic condition than has hydrogen, since the hydrogen ions present in dilute sulphuric acid so readily pass on their charges to zinc, themselves becoming discharged. This tendency is a measurable quantity which can be expressed in volts.

The table below, in which some of the better known

metals are arranged in the order of their reactivity as de
scribed on p. 166, also shews their potentials with reference
to hydrogen, when in contact with solutions of their respective salts of normal ionic concentration. The con-
vention as to the use of + and − signs, now usually adopted
in such tables, is to take hydrogen as zero. Metals which are
capable of liberating hydrogen from dilute acids are assigned a negative potential, because they function in
general as the negative electrode in voltaic cells, while those
metals which do not, under ordinary conditions, liberate
hydrogen from dilute acids, and which likewise function as
the positive electrode in such cells, are regarded as positive*.

Metal	Electrode potential (volt)	Metal	Electrode potential (volt)
Potassium	−2·9	Lead	−0·132
Sodium	−2·7	Hydrogen	0
Calcium	−1·9	Copper (against cupric)	+0·347
Magnesium	−1·8	Mercury (against mer-	+0·793
Aluminium	−1·337	curous)	
Zinc	−0·77	Silver	+0·799
Iron (against ferrous)	−0·43	Platinum	+0·86
Tin (against stannous)	−0·146	Gold	+1·08

The electrode potential order of the metals is in general
identical with the displacement order as described above,
but in certain cases this can be modified by alteration in
the concentrations of the solutions. Lead follows tin in the
table, but each can displace the other from solutions of

* In cells of constant electromotive force, such as that of Daniell,
polarization is obviated. An approximate value of the electro-motive
force of such a cell, which consists of the combination Zn : ZnSO₄ :
CuSO₄ : Cu can be calculated from the table. Adding the values of the
two normal potentials, the algebraic sum equals 1·117 volts. The actual
value is about 1·1 volts.

their respective salts by appropriate adjustment of the concentration. For instance, if lead foil be immersed in a solution of stannous chloride, preferably acidified with hydrochloric acid to avoid the separation of basic salt, lead goes into solution and tin is deposited. On the other hand, if tin foil be immersed in a hot solution of lead chloride, lead is precipitated and stannous chloride is found in solution.

A consideration of the table shews that the order of electro-chemical potential of the metals there enumerated follows that of their general chemical reactivity, especially as regards their avidity for oxygen. The alkali metals oxidize rapidly in air at ordinary temperatures, and therefore are usually kept under hydrocarbon oils. They decompose water at ordinary temperatures, with evolution of hydrogen and production of the hydroxide. A newly-cut surface of sodium or potassium rapidly acquires a coat of deliquescent hydroxide which soon absorbs carbon dioxide from the atmosphere, finally forming a white friable crust of carbonate. In dry air the metals tarnish more slowly, forming first the oxide and ultimately the carbonate. The salts formed by these metals with strong acids have a neutral reaction, since they do not hydrolyse appreciably.

The alkaline earth metals calcium, strontium, and barium somewhat resemble the alkali metals in these respects. They oxidize slowly in air at ordinary temperatures, and their salts with strong acids shew little tendency to hydrolyse. Magnesium, which belongs to this group, is somewhat abnormal, since it does not act appreciably on cold water, tarnishes but little in air, and yields salts which hydrolyse slightly.

Metals such as aluminium and zinc are in a different category. Their oxides are much less strongly basic, as is shewn by the hydrolysis of their salts. Thus, solutions of aluminium and zinc chlorides have a strongly acidic reaction. The amphoteric character of the oxides, i.e. their power of reacting both as acidic and as basic oxides (*vide infra*), is further seen by the metals dissolving in caustic alkalis with evolution of hydrogen, and the formation of the alkali aluminate or zincate (p. 58). The oxides of tin are also amphoteric, reacting as basic oxides towards strong acids, and as acidic oxides towards the caustic alkalis.

Some metals, notably those of the alkali and alkaline earth series, give rise to only one series of salts. Others, such as copper, iron, and mercury, give two. When a metal forms more than one oxide there may be a series of salts corresponding to each oxide, but not necessarily so. Thus, iron forms three well-defined oxides, ferrous oxide, FeO, ferric oxide, Fe_2O_3, and magnetic oxide, Fe_3O_4. Ferrous oxide produces ferrous salts, ferric oxide yields ferric salts, whereas magnetic oxide, a compound oxide of the other two, gives rise to a mixture of ferrous and ferric salts. On the other hand, while there are two oxides of barium, the only series of barium salts corresponds to the monoxide, BaO.

The increase in valency (ionic charge) of the metal, brought about by the oxidation of an -*ous* salt to an -*ic* salt, results in a weakening of basic properties. Thus ferric oxide is a weaker base than ferrous oxide, as is shewn by the considerably greater hydrolysis of the ferric salts.

Although the metals are more numerous than the non-metals, the chemistry of their compounds is, for the greater part, much simpler. The ionic hypothesis has greatly facilitated the study of metallic compounds, and in general, when the properties of the oxides are known, the characteristics of the salts derived from them can be deduced.

Properties of Metals and Alloys. The properties of metals depend greatly upon the method of preparation and subsequent treatment. For instance, when precipitated from solution by more electro-positive metals, they are usually obtained in finely divided condition. Copper precipitated from copper sulphate solution by zinc appears as a very dark powder. Some metals are very brittle at temperatures only slightly below their freezing points. Zinc dust and lead powder can be made by grinding the respective metals under these conditions. The specific gravity of cast metals is raised by mechanical working, such as hammering, rolling, and drawing, these operations usually hardening the metal considerably. The process of annealing, i.e. heating, softens the metal again, rendering it more workable and less brittle. The tensile strength and hardness of copper and certain other metals is greatly increased by cold drawing.

The less electro-positive metals, such as silver and copper, can be precipitated in mirror form on a glass surface by reduction of their solutions under suitable conditions. If a little aqueous caustic soda be added to silver nitrate solution, and the resulting silver oxide nearly dissolved by dilute ammonia, the addition of a solution of grape sugar and subsequent warming will precipitate mirror silver on the walls of the containing vessel. Copper mirrors are more difficult to obtain, and special reducing agents are required. If to a 0·5 per cent. solution of copper acetate, dilute ammonia be added until the precipitate at first formed is just redissolved, and then a solution of hydrazine until the colour is discharged, on warming in a water bath a copper mirror appears on the walls of the containing vessel.

Like other elements, many metals exhibit the phenomenon of allotropy. Perhaps the most striking instance of this at present known is that of tin, though allotropy is probably commoner among metals than has hitherto been supposed. Tin is usually known as a lustrous white metal with a very marked crystalline structure, but when kept for a long time at low temperatures it changes to a grey powder. This is an enantiotropic change, since the transition temperature is well marked at 18° C. Ordinary white tin is the stable phase above 18° C., while grey tin is stable below this temperature. As the transformation of white into grey tin takes place very slowly, objects made of this metal can withstand temperatures below the ordinary for a long time without appreciable change. Nevertheless old coins containing tin frequently exhibit signs of deterioration, the phenomenon being known as the tin plague. Antimony is usually a white, brittle, and highly crystalline metal. When obtained by electro-deposition from a solution of the chloride, the product on the surface of the cathode is very unstable. A slight scratch is sufficient to cause a violent change to ordinary antimony. Since this change is irreversible, these forms of antimony are generally regarded as furnishing an instance of monotropic allotropy. Iron, on cooling from high temperatures, also shews changes which are interpreted as evidence of the existence of allotropes.

The presence of even traces of impurity sometimes exercises an enormously modifying influence upon the physical properties of metals. Minute traces of arsenic in copper diminish its electric conductivity very seriously. Lord Kelvin pointed out in 1860 that the presence of 0·1 per cent. of bismuth in copper would reduce its conductivity to such an extent as to be fatal to the success of the Atlantic cable, if such material were used. The addition of 0·2 per cent. of bismuth to gold would, from the point of view of coinage, convert the gold into a useless material which would crumble under the pressure of the die, and traces of lead behave similarly.

As the properties of metals are much modified by admixture with others, various technical purposes are better secured by alloys than by pure metals. Indeed the practical preparation and use of alloys has in many instances preceded their scientific study, as in the case of such well-known alloys as brass (copper and zinc) and various combinations of copper and tin, such as bronze and bell metal.

A few pairs of metals are almost immiscible in the liquid state, such as aluminium and lead, and lead and zinc (cf. Parkes's process, p. 284), but the greater number are soluble in each other, if not in all proportions, yet very nearly so when liquid, forming homogeneous melts. In such products the metals may be chemically combined, or may be merely dissolved in each other. The condition of metals in an alloy is determined by studying the freezing point curves of mixtures of varying composition, supplemented with microscopic examination of the products. When definite compounds are not formed, the constituents of the solid alloy may be present either as solid solutions (mixed crystals) or as conglomerates (separate crystals). There are also more complex possibilities. A compound may, for example, form mixed crystals with excess of either or both of its constituents.

When mercury is one of the components of an alloy the product is known as an amalgam, but this does not differ in any essential from other alloys. Amalgams are characterized by their low melting points, but alloys of other metals, particularly tin, lead, bismuth, and cadmium, also possess remarkably low melting points. Several ternary alloys of these metals are known which melt below the temperature of boiling water. An alloy of sodium and potassium, resembling mercury in appearance but lighter than water, is liquid at ordinary temperatures, though its application is very restricted on account of the chemical activity of its constituents.

When an alloy is treated with an acid which is capable

of dissolving its separate constituents, the acid sometimes exercises a selective solvent action, in so far as the more electro-positive metal is the first to be attacked. Lead-silver alloys, for instance, when treated with nitric acid, part with practically all the lead before the silver is attacked, or, if the silver is taken into solution it is displaced again by the lead. This is of obvious importance in analysis; the whole of an alloy taken must be dissolved prior to examination, otherwise there can be no certainty that all the metals are present in solution.

Metallic Hydroxides and Oxides. The hydroxides of sodium and potassium are manufactured on a very large scale at the present time by electrolytic methods. Chlorine and sodium hydroxide are produced simultaneously by the electrolysis of brine. The sodium ions which form the sodium hydroxide must be separated from the brine before discharge. This is done by using divided cells, and many devices are employed for this purpose, such as porous partitions and mercury septa. On a small scale the preparation can be illustrated by immersing a porous pot, containing pure water to which a few drops of soda have been added to render it conductive, in an outer vessel of brine. An iron rod will serve as cathode in the inner cell, and a rod of carbon as anode in the brine. On passage of a current chlorine is evolved at the anode. The concentration of sodium hydroxide in the cathode compartment continuously increases, as can be shewn by withdrawing measured volumes of the liquid by a pipette from time to time and titrating with standard acid. On the manufacturing scale, the solution is evaporated down and the fused soda cast into sticks or blocks.

Before the advent of electrolytic methods, sodium hydroxide was manufactured from the carbonate by the process known as causticizing with lime. When a dilute solution of sodium carbonate is boiled with calcium hydroxide in calculated quantities, insoluble calcium carbonate separates and a solution of sodium hydroxide is obtained:

$$Na_2CO_3 + Ca(OH)_2 \rightleftharpoons CaCO_3 + 2NaOH.$$

As the reaction can reverse to some extent in concentrated solution, the liquid must be kept dilute. The boiling is continued until the filtered liquid gives no reaction for carbonate. It is then allowed to settle, the clear solution withdrawn and concentrated by evaporation to the required degree. The product so obtained is sufficiently pure for soap making.

Sodium hydroxide is deliquescent, and therefore very soluble in water. Both the solid and the aqueous solution rapidly absorb carbon dioxide, forming the carbonate. When added to solutions of metallic salts the metallic hydroxide is generally precipitated, since this class of compound is marked by great insolubility. Thus with ferric chloride a bulky reddish brown precipitate of ferric hydroxide is produced. Copper sulphate gives a light blue precipitate of cupric hydroxide:

$$FeCl_3 + 3NaOH = Fe(OH)_3 + 3NaCl,$$
$$CuSO_4 + 2NaOH = Cu(OH)_2 + Na_2SO_4.$$

Sometimes the precipitated hydroxide re-dissolves when excess of the alkaline hydroxide is added. Such are the amphoteric hydroxides of aluminium, lead, zinc, and tin (both stannous and stannic):

$$AlCl_3 + 3NaOH = Al(OH)_3 + 3NaCl,$$

and on the further addition of soda, soluble sodium aluminate is formed:

$$Al(OH)_3 + NaOH = NaAlO_2 + 2H_2O,$$

in which the aluminium is now a part of the acid radical, or anion.

In certain cases it is the *oxide* which is precipitated, and not the hydroxide. The salts of silver, mercury, and copper (cuprous) react thus, and also cupric salts when hot solutions are used. Practically all the metallic hydroxides, except those of the alkali metals, lose water on heating. In some cases this is a reversible reaction, but not usually. Possibly the temperature at which the hydroxides of silver and mercury resolve into oxide and water is below the freezing point of the solutions. There is indeed some evidence for this, but it is not important, since in but few instances are metallic hydroxides definite compounds of the oxide and water. They are better regarded as adsorption products of the type of silicic acid. This is shewn by the effect of gradually withdrawing water from moist aluminium hydroxide, when the vapour pressure curve is continuous, shewing no break at any point. That is, there is no discontinuity in the curve when more water is lost than the formula $Al(OH)_3$ requires. Such a substance is more accurately represented as $Al_2O_3 . xH_2O$ and more correctly described as hydrated aluminium oxide.

Many of these metallic oxides and hydroxides can also be obtained by the action of ammonia upon the salts. The precipitates are for the greater part identical with those produced by the action of soda on the solution (or in some cases, upon the solid salt). When excess of ammonia is added, however, variations frequently appear. For example, cupric salts with ammonia yield cupric hydroxide, but a slight excess of the reagent readily redissolves the precipitate, forming a deep blue solution of a complex cuprammonium salt. Salts of mercury are very irregular. In no instance does the addition of ammonia to a mercurous or mercuric salt, solid or in solution, yield either the oxide or the hydroxide. The products are of a complex character. Metals of a less basic type yield hydroxides less soluble in excess of ammonia than in caustic alkalis.

Since metallic hydroxides, with the exception of those of sodium and potassium, lose water on ignition, leaving the oxide, and since most hydroxides can be obtained by precipitation, this forms a very general method for the preparation of basic oxides. Other general methods, such as ignition of the nitrate or carbonate of the metal, have already been described. Many basic oxides can conveniently be prepared by heating the metal in air or in oxygen, and this is sometimes the commercial process of manufacture. Litharge, PbO, and magnetic iron oxide, Fe_3O_4, are exclusively manufactured in this way. Sometimes this method yields a higher oxide. Sodium, for instance, when heated in dry air, first yields the (impure) basic oxide, Na_2O, but on continued ignition, and especially when oxygen is used, gives the peroxide, Na_2O_2, a yellowish powder of strongly oxidizing properties which yields hydrogen peroxide when in contact with water. Potassium peroxide, K_2O_4, has very similar properties and is made in the same way. Litharge is also further oxidized to red lead, Pb_3O_4, by heating in air; indeed, this is the only practicable way of obtaining this compound.

When a basic oxide is dissolved or suspended in sodium hydroxide solution and chlorine passed through the liquid, a higher oxide is generally produced. The dioxides of lead and manganese, PbO_2 and MnO_2, are conveniently prepared in this way. A modification of this method consists in the addition of excess of sodium hypochlorite to a solution of the salt. On boiling or standing, the same products are usually produced.

In the electrolysis of certain salts the anodic products are powerfully oxidizing in their action. It has already been seen that potassium persulphate and other highly oxidized products are formed in this way. If two lead plates are immersed in sulphuric acid of about 1·2 sp. gr., and a current of electricity be sent through the circuit, hydrogen is evolved at the cathode, while the anodic products attack the metal, which becomes coated with an adherent film of lead dioxide. If the external source of current be removed, the lead plates will be found capable of yielding a current themselves, the plate coated with dioxide acting as the positive electrode. This is the process of charging and discharging an ordinary lead accumulator. During the charging operation the concentration of the sulphuric acid increases, but falls as the cell is discharged. The reactions involved in charging and discharging the cell are usually expressed thus:

$$\text{In charging} \xrightarrow{\hspace{3cm}}$$
$$2PbSO_4 + 2H_2O \rightleftharpoons PbO_2 + Pb + 2H_2SO_4,$$
$$\xleftarrow{\hspace{3cm}} \text{In discharging}$$

but various objections to this explanation can be raised, and certainly the equation is not quantitative. Suggestions of higher oxides of lead have been introduced in order to explain the mechanism of the reactions more in accordance with observed facts, but the chemistry of lead storage cells is complex and difficult, and theories concerning their action are still controversial.

A study of the specific properties of the derivatives of any metal begins with a consideration of the properties of the oxides and of such ions as may be derived from them. Certain metallic oxides may now be studied from this standpoint.

Oxides and Ions of Iron. Ferrous oxide, FeO, is difficult to prepare in pure condition on account of the ease with which it undergoes further oxidation. It can be obtained by gently heating ferrous oxalate, which loses carbon monoxide and dioxide:

$$FeC_2O_4 = FeO + CO + CO_2.$$

The residue is a black powder, so finely divided that when scattered into the air it combines with oxygen with incandescence. The hydroxide precipitated from solutions of ferrous salts is almost colourless at first, but rapidly oxidizes. Ferrous oxide cannot therefore be prepared in pure condition by the general method of heating the hydroxide.

The salts corresponding to this oxide, which however are more conveniently prepared from the metal, have a pale green colour, most of them crystallizing with water of hydration. The ferrous ion, $\overset{++}{Fe}$, is almost colourless and is marked by its readiness to acquire another unit of positive charge, becoming $\overset{+++}{Fe}$, the ferric ion. Ferrous salts are thus strong reducing agents. Gold may be precipitated in metallic condition from solutions of any of its salts by the addition of ferrous sulphate.

Ferric oxide is a natural product of very wide occurrence. It is the chief ore from which iron is obtained in this country, and is the main constituent of iron rust. Its wide dissemination in small quantities can be seen in the almost universal occurrence of its colour in clay, sand, and rocks; it is indeed Nature's most abundant pigment.

In pure condition it is usually prepared either by heating crystallized ferrous sulphate, when the residue is known as jewellers' rouge and finds application in the polishing of both metals and glass, or by dehydrating precipitated ferric hydroxide. When heated to very high temperatures, ferric oxide suffers appreciable loss of oxygen, with partial conversion into magnetic oxide. This latter is a compound oxide which should be written $FeO . Fe_2O_3$, since when heated with hydrochloric acid both ferrous and ferric chlorides are obtained, in the proportions represented by the equation

$$Fe_3O_4 + 8HCl = 4H_2O + FeCl_2 + 2FeCl_3.$$

The ferric ion is almost colourless, the high colour of ferric chloride solutions being largely due to colloidal ferric hydroxide produced by hydrolysis. If a few drops of dilute sulphuric acid be added to a solution of ferric chloride, the deep colour is almost discharged. The sulphuric acid has suppressed hydrolysis, and therefore removed the colour due to colloidal ferric hydroxide. The addition of concentrated hydrochloric acid to ferric chloride solution slightly changes the tint to a golden brown. Hydrolysis is suppressed as before, but so also is the ionization of the ferric chloride, giving a preponderance to the undissociated molecules of the latter, which do not differ greatly in colour from colloidal ferric hydroxide.

In those cases where ferrous compounds do not shew a strong tendency to pass into ferric; that is, where the ferrous compound is the more stable, it will be found that the simple ferrous ion, $\overset{++}{Fe}$, does not exist. For example, ferrocyanides (which may be regarded as derived from ferrous salts) are more stable than ferricyanides (derived from ferric salts), since crystals or solutions of the latter somewhat readily tend to pass into ferrocyanide, while ferrocyanides are not oxidized by air into ferricyanides. As has been seen, these compounds do not contain the simple ions of iron, the iron being part of the complex anions.

Oxides and Ions of Tin. Tin forms two oxides, stannous oxide, SnO, and stannic oxide, SnO_2, each having a corresponding series of salts. Stannous oxide can be obtained by heating stannous oxalate. The oxide itself is unimportant, and the salts corresponding to it are more conveniently prepared from the metal (cf. iron). Stannic oxide occurs as tinstone, the commonest ore of the metal. It is largely used for the production of frits, enamels, and glazes. As in the case of iron, the ion with the higher valency is the more stable. The stannous ion, $\overset{++}{\text{Sn}}$, owes its powerful reducing properties to its readiness to acquire two additional units of charge. Both the ions of tin are colourless, and their salts are considerably hydrolysed in solution.

Oxides and Ions of Lead. The best known oxides of lead are litharge or massicot, PbO, red lead or minium, Pb_3O_4, and the dioxide, PbO_2, all of which shew basic properties in a greater or smaller degree.

Litharge is obtained commercially almost entirely as a bye-product in the cupellation of lead for the removal of silver. The oxide is slightly soluble in water, to which it imparts a faintly alkaline reaction. As a laboratory experiment, it can be prepared by any of the general methods described above. It is the most stable of the oxides of lead, since all other oxides of this metal yield litharge when heated to fusion. Its basic character is very pronounced. When fused with silica it yields a heavy, low melting glass, and for this reason lead oxides should not be heated strongly in siliceous apparatus. It has in the past been used for glazing certain kinds of earthenware, but such glazes have the disadvantage of being slightly soluble and very poisonous. Practically all the usual salts of lead are derived from the monoxide.

When heated in air to a temperature just below redness, powdered litharge is slowly oxidized to red lead, Pb_3O_4. The process requires several days, and is continued until the product has a fine scarlet colour. When red lead is warmed with dilute nitric acid a solution of lead nitrate is obtained and an insoluble residue of the puce-coloured dioxide, PbO_2, is left:

$$Pb_3O_4 + 4HNO_3 = 2Pb(NO_3)_2 + PbO_2 + 2H_2O.$$

The mass of lead found in solution is twice that of the lead left as dioxide. Red lead is therefore a compound oxide of composition represented by the formula $PbO_2 . 2PbO$. Although red lead and magnetic oxide can both be represented by the formula M_3O_4, it will be seen that they are of quite different types. This is confirmed by boiling them with concentrated hydrochloric acid, when chlorine is evolved from red lead. No chlorine can be obtained thus from magnetic iron oxide. Red lead is used in the making of paint, and as a kind of cement for the joints of iron pipes.

Methods for preparing lead dioxide have already been given. Its basic properties are very feeble, but it can give rise to a corresponding chloride, $PbCl_4$, and certain other salts such as the tetracetate. These salts of quadrivalent lead are, however, very unstable. When heated with concentrated hydrochloric or sulphuric acid, lead dioxide yields salts of bivalent lead, together with chlorine from the former acid, and oxygen from the latter, a reaction very common among higher metallic oxides:

$$PbO_2 + 4HCl = PbCl_2 + 2H_2O + Cl_2,$$

$$2PbO_2 + 2H_2SO_4 = 2PbSO_4 + 2H_2O + O_2.$$

Since lead dioxide is a constituent of red lead, these reactions, as might be expected, are also given in a modified degree by the latter compound.

$$Pb_3O_4 + 8HCl = 3PbCl_2 + 4H_2O + Cl_2.$$

The quadrivalent plumbic ion, $\overset{++++}{Pb}$, is unstable and easily passes into the bivalent ion, $\overset{++}{Pb}$, the salts of which in general shew only slight solubility.

White Lead. One of the most important salts of lead from the technical point of view is the basic carbonate or white lead, which approximates in composition to the formula $2PbCO_3 . Pb(OH)_2$. This is the basis of the common kind of oil paint, and its value depends very largely on its physical condition. Although many quite simple methods of making it are known, the cheaper products have not quite the same value as regards 'body' or covering power as the material made by the older manufacture, known as the Dutch process. The newer and quicker methods produce a white lead which is more crystalline, and therefore less opaque. In the Dutch process, lead cast into gratings of a suitable size and shape is placed in pots containing a little ordinary vinegar or dilute acetic acid. These pots are then arranged in heaps, the successive layers being covered with loose boards on which stable refuse or spent tan is spread, with the object of maintaining a moderate temperature and supplying carbon dioxide by its fermentation. The pots are left undisturbed for some weeks after which the heap is dismantled, the white crust of basic carbonate scraped away, washed to remove soluble matter, and finely ground.

The chemical changes which take place may be summarized thus:

Lead acetate + lead + air → basic lead acetate.

Basic lead acetate + carbon dioxide → basic lead carbonate + normal lead acetate,

the process being thus continuous.

Substitutes for white lead have been sought, partly on account of cost, and partly because of the slow blackening which paints derived from white lead undergo in presence of traces of hydrogen sulphide. The chief of the substitutes are barium sulphate, lead sulphate, and zinc oxide. Barium sulphate does not tend to darken, but, on the other hand, has but poor 'body' and is liable to blister or peel away from the surface. Basic lead sulphate is a very common substitute. It is more crystalline than white lead, but does not darken. For many purposes zinc oxide is superior even to white lead, since it alters but little under atmospheric influences, has good 'body,' and is very durable.

Oxides and Ions of Copper. Copper forms two well-defined oxides, cuprous oxide, Cu_2O, and cupric oxide, CuO, both of which are basic and yield corresponding salts. Those derived from the higher oxide are usually the more stable and are for the greater part soluble in water.

Cupric oxide can be prepared by any of the standard methods, the purest form, such as is used in organic analysis, being obtained by heating copper wire in air or oxygen. Since it is readily reduced at a red heat it is a useful oxidizing agent.

Cuprous oxide occurs native to a small extent in crystalline form as ruby copper ore. When copper is heated in air both oxides are at first formed, the final product at a red heat being cupric oxide, though at the highest temperatures loss of oxygen is perceptible. The following experiment illustrates a convenient method of small scale preparation of cuprous oxide.

A few grammes of cupric oxide and excess of copper turnings are boiled with concentrated hydrochloric acid until the colour has almost disappeared. The clear liquid can be poured into water from the excess of copper, when a copious precipitate of colourless cuprous chloride separates.

$$Cu + CuO + 2HCl = 2CuCl + H_2O.$$

This settles on standing, and the supernatant liquid can be poured away. If excess of caustic soda solution be added to the milky residue, cuprous oxide is formed:

$$2CuCl + 2NaOH = Cu_2O + 2NaCl + H_2O.$$

As a variation the cuprous chloride can be dissolved in ammonia, in which it is very soluble, and soda at once added, since oxidation in ammonia solution proceeds with great rapidity. As has been seen, cuprous chloride is insoluble in water but dissolves in ammonia and in concentrated hydrochloric acid, these being practically the only solvents. Both solutions are used as absorbents for carbon monoxide in gas analysis. Cuprous chloride is unstable in presence of air, and is usually kept as a solution in hydrochloric acid in presence of a strip of copper to prevent oxidation.

Cuprous oxide can be obtained by many other processes of reduction. Cupric hydroxide dissolves readily in solutions of organic compounds rich in hydroxyl groups, such as glycerine, sugars, tartrates, etc. Fehling's solution is essentially cupric hydroxide dissolved in Rochelle salt (potassium sodium tartrate) and is a well-known 'test' for reducing agents. On boiling with grape sugar for instance, Fehling's solution yields a precipitate of cuprous oxide, the reduction not proceeding further.

The deep blue solutions which cupric hydroxide yields with organic compounds containing OH— groups may be regarded as cupric hydroxide in which the hydrogen atoms have been replaced by organic complexes. The copper in these compounds is part of a complex anion. It does not exist as the simple bivalent cupric ion, $\overset{++}{\text{Cu}}$, as is shewn by electrolysing Fehling's solution in a voltameter in series with one containing copper sulphate solution. In the copper sulphate the blue colour, which of course is due to cupric ions, can be seen to travel towards the cathode, while in the Fehling's solution the blue colour wanders in the opposite direction, towards the anode.

Cuprous salts shew nothing of the characteristic blue colour of cupric salts. The cuprous ion is considered to be colourless, though few cuprous salts dissolve in water. Their solutions in such liquids as hydrochloric acid and ammonia, although colourless, give little indication as to the colour of the cuprous ion itself.

Oxides and Ions of Mercury. Mercury forms two oxides, mercurous, Hg_2O, and mercuric, HgO, each giving rise to a corresponding series of salts.

The preparation of mercuric oxide by oxidation of the metal in air, from which oxygen was first obtained by Priestley, formed part of the classical researches of Lavoisier, but this does not offer a convenient method of preparing the compound. The process is very slow and only small quantities can be got in this way. It is better prepared by the general method of heating the nitrate (obtained by the action of excess of nitric acid on the metal) which gives *red* oxide of mercury,

$$2Hg(NO_3)_2 = 2HgO + 4NO_2 + O_2,$$

this decomposition taking place at a lower temperature than is required for the dissociation of the oxide. A *yellow* modification is given by the action of caustic soda on mercuric salts, in solution or otherwise. Much discussion has taken place as to whether the red and yellow varieties are to be considered as polymorphic forms, or whether the difference in colour is due to difference in the size of the particles. The evidence tends to favour the latter view. Both forms become black when heated just short of the temperature of dissociation, but each returns to its original colour on cooling. Mercuric salts can be obtained from this oxide by the action of the corresponding acids.

Mercurous oxide is yielded by the action of caustic soda on mercurous salts, solid or in solution. It is easily obtained as a heavy black powder by warming calomel, $HgCl$, with excess of sodium hydroxide solution, and gives the corresponding salts (mercurous) when treated with acids.

When mercury is treated with nitric or concentrated

sulphuric acid, excess of the metal yields mercurous salts,
whereas excess of the acid yields mercuric salts. Since
hydrochloric acid does not attack mercury, the chlorides
are very generally prepared from mercuric sulphate. The
operation consists in grinding together solid mercuric sulphate and common salt. When mercuric chloride is required, a trace of some oxidizing agent, such as manganese
dioxide, is added to avoid any reduction taking place.
When mercurous chloride is sought, the common salt and
mercuric sulphate are ground with the calculated weight
of the metal. In both cases the intimate mixture is heated
in dry retorts, when the required chloride volatilizes and
is condensed in the colder parts of the apparatus:

$$HgSO_4 + 2NaCl = Na_2SO_4 + HgCl_2,$$

$$HgSO_4 + Hg + 2NaCl = Na_2SO_4 + 2HgCl.$$

Purification of the product is easily carried out, since
mercuric chloride is soluble in water while mercurous
chloride is insoluble. Careful purification is essential, since
mercuric chloride (corrosive sublimate) is an active poison.
It possesses remarkable antiseptic, astringent, and preservative properties, while calomel is extensively used
medicinally.

When a solution of potassium iodide is added to a solution of mercuric chloride a precipitate of mercuric iodide
is thrown down. This dissolves in excess of either solution.
Nessler's reagent, used for detecting and estimating traces
of ammonia, is made by dissolving mercuric iodide in
potassium iodide solution in presence of excess of potash.

The two modifications of mercuric iodide, the scarlet
and the yellow, are respectively stable below and above the
transition temperature 126° C. The compound is therefore
enantiotropic. When heated above 126° C. and allowed
to cool, the metastable yellow form does not immediately

revert to the scarlet modification, but can be made to do so by gentle friction. This is usually shewn by rubbing the scarlet compound on a filter paper, and heating gently until the yellow modification is produced. When cold, the yellow form returns to the scarlet wherever it is touched with a glass rod. When the compound is precipitated from solution the unstable modification is the first to be formed, as would be expected. By using very dilute and cold solutions of potassium iodide and mercuric chloride the change from yellow to scarlet can be retarded sufficiently to be easily visible. If solutions in dilute alcohol be used, the rate of change is even further retarded. When required in quantity, mercuric iodide is made by rubbing together the calculated weights of mercury and iodine moistened with alcohol.

Mercuric salts are very little ionized in solution, as is shewn by experiments on their osmotic effects and electrical conductivity. Abnormalities in reaction are therefore to be expected. When mercuric chloride is covered with concentrated sulphuric acid and heated, the chloride melts to a colourless liquid below the acid, and ultimately volatilizes, no hydrogen chloride being evolved. If a little mercuric oxide (either red or yellow) be shaken with water and phenol-phthalein added, no colour appears in the solution. On the addition of sodium chloride the liquid shews an alkaline reaction, due to the sparing ionization of the mercuric chloride:

$$HgO + 2\overset{+}{Na} + 2\overset{-}{Cl} + H_2O \rightleftarrows HgCl_2 + 2\overset{+}{Na} + 2(\overset{-}{OH}).$$

Potassium permanganate and potassium dichromate are of great importance and will be considered separately.

Potassium Permanganate has for long been manufactured by fusing potash and pyrolusite (native manganese dioxide) together in presence of air, when potassium manganate is formed:

$$4KOH + 2MnO_2 + O_2 = 2K_2MnO_4 + 2H_2O.$$

The process is hastened and the yield increased by using oxidizing substances such as potassium chlorate or nitrate, but this is not a practicable method commercially. The green-coloured melt is extracted with water, filtered from unchanged manganese dioxide, and converted into permanganate. This can be accomplished in several ways. Potassium manganate undergoes hydrolysis in solution with formation of permanganate:

$$3K_2MnO_4 + 2H_2O = 2KMnO_4 + MnO_2 + 4KOH.$$

This is an instance of a single compound undergoing simultaneous oxidation and reduction, part of it being oxidized to permanganate at the expense of another part, which is reduced to manganese dioxide, the phenomenon being known as *autoxidation*. Dilution with water therefore converts manganates into permanganates. The same conversion is more conveniently brought about by the action of carbon dioxide or acids:

$$3K_2MnO_4 + 4CO_2 + 2H_2O = 2KMnO_4 + 4KHCO_3 + MnO_2,$$

and this is the process usually adopted, since in most cases it is cheaper to sacrifice part of the permanganate and refuse the precipitated manganese dioxide with potash. Oxidation with chlorine or ozone gives a better yield, since no manganese dioxide is precipitated:

$$2K_2MnO_4 + Cl_2 = 2KMnO_4 + 2KCl,$$

$$2K_2MnO_4 + O_3 + H_2O = 2KMnO_4 + O_2 + 2KOH.$$

In practice sodium hydroxide is used instead of the more costly potash, and the less easily crystallizable sodium permanganate converted into the potassium salt by double decomposition with the calculated quantity of potassium chloride, the relative solubilities of the compounds concerned rendering this an easy process.

If dilute potash be electrolysed in a divided cell, using an anode of iron-manganese alloy (and pig iron containing a high percentage of manganese is easily produced) potassium permanganate is formed by anodic oxidation. No iron passes into solution, but much ferric oxide separates out. This process is used on a manufacturing scale.

Potassium permanganate is not very soluble, a saturated solution at laboratory temperatures containing only about 5 per cent. of the salt. Such a solution has so deep a colour as to be almost opaque. When a large excess of potash is added to a dilute solution of potassium permanganate, green manganate is formed with liberation of oxygen, but the volume of gas is relatively so small that it remains dissolved in the liquid and escapes observation. If the salt be boiled with excess of potash the oxygen can be collected:

$$4KMnO_4 + 4KOH = 4K_2MnO_4 + 2H_2O + O_2.$$

Like most strongly oxidizing compounds, permanganates yield chlorine when treated with hydrochloric acid, and oxygen when warmed with sulphuric acid. When solid potassium permanganate is gently warmed with a little concentrated sulphuric acid, a violent reaction takes place, the anhydride of permanganic acid is at first formed, but this is very unstable and decomposes with evolution of oxygen and a separation of manganese dioxide.

The ease with which permanganates yield oxygen to oxidizable substances such as ferrous salts and oxalates, even in dilute solution, together with the complete discharge of the colour when this takes place, renders it very useful for quantitative estimations. In presence of dilute sulphuric acid the reaction follows the equation:

$$2KMnO_4 + 3H_2SO_4 = K_2SO_4 + 2MnSO_4 + 3H_2O + 5O.$$

From this it will be seen that 316 grm. of potassium permanganate yield $5 \times 16 = 80$ grm. of 'available' oxygen. It should be noted that in this case the oxygen is not liberated *qua* gas, since the solutions used are very dilute.

As 31·6 grm. of potassium permanganate furnish 8 grm. of available oxygen, the equivalent weight of the salt is 31·6 when used under these conditions, i.e. in presence of excess of dilute sulphuric acid.

In presence of alkalis, potassium permanganate is a most useful oxidizing agent, especially towards organic compounds. The equivalent weight is larger in this case, since the reaction proceeds thus:

$$2KMnO_4 + H_2O = 2KOH + 2MnO_2 + 3O.$$

That is, 316 grm. of potassium permanganate yield 48 grm. of available oxygen, and the equivalent weight

$$= \frac{316 \times 8}{48} = 52·7.$$

Potassium Dichromate, like all other chromium derivatives, is manufactured from chrome iron ore, a compound oxide of iron and chromium of the formula

$$(FeO)_x . (Cr_2O_3)_y.$$

The ore is heated to redness and suddenly quenched in water, after which treatment it can readily be ground to powder. This is mixed with potassium carbonate and limestone, the latter being added to prevent fusion, and oxidized by roasting in air. Potassium nitrate is sometimes added to hasten the oxidation. The ferrous oxide is converted into ferric oxide, and the oxide of chromium into potassium chromate:

$$2Cr_2O_3 + 4K_2CO_3 + 3O_2 = 4K_2CrO_4 + 4CO_2.$$

The chromate is obtained in solution by lixiviating the mass with water, and drawing off from the insoluble residue.

On adding a slight excess of acid, the chromate is converted into dichromate:

$$2K_2CrO_4 + 2HNO_3 = 2KNO_3 + K_2Cr_2O_7 + H_2O,$$

which crystallizes out on evaporation. As in the case of permanganates, the sodium salt is generally the first to be prepared, for economic reasons, the potassium salt being obtained from this by double decomposition with potassium chloride.

If concentrated sulphuric acid be cautiously and slowly added to about an equal bulk of a warm saturated solution of potassium dichromate and the mixture allowed to cool, fine crimson needles of chromic anhydride, CrO_3, separate. These can be filtered off by using a loose plug of asbestos, washed with concentrated nitric acid to remove sulphuric acid, and dried on a porous earthenware plate. Crystals of chromic anhydride are deliquescent and very soluble in water, yielding a solution called chromic acid, though no definite compound corresponding to H_2CrO_4 has been obtained. The elements sulphur and chromium are closely related, and there is much similarity both between their acids and the salts derived from them. Potassium chromate is isomorphous with potassium sulphate, though potassium bichromate is analogous, not to potassium bisulphate, $KHSO_4$, but to the pyrosulphate, $K_2S_2O_7$. *True* potassium bichromate would be $KHCrO_4$, but no such compound can be prepared. Water is at once lost:

$$2KHCrO_4 = K_2Cr_2O_7 + H_2O.$$

Solutions of dichromates have an acid reaction and a deep orange colour. On the addition of alkalis this changes to bright yellow. The anion of dichromates differs from that of chromates therefore, the former being $\overset{--}{Cr_2O_7}$, and the latter $\overset{--}{CrO_4}$, though the conversion of the one into the other is readily brought about by the addition of alkali or acid respectively. Thus, chromates in presence of acids yield bichromates (orange):

$$2K_2CrO_4 + H_2SO_4 = K_2SO_4 + K_2Cr_2O_7 + H_2O,$$

and bichromates with alkalis yield chromates (yellow):

$$K_2Cr_2O_7 + 2KOH = 2K_2CrO_4 + H_2O.$$

Bichromates are very generally used as oxidizing agents, both for organic and inorganic compounds. In presence of dilute acid they readily yield oxygen to oxidizable substances, such as ferrous salts and alcohols. The reaction follows the equation:

$$K_2Cr_2O_7 + 4H_2SO_4 = K_2SO_4 + Cr_2(SO_4)_3 + 4H_2O + 3O,$$

that is, each molecule of potassium bichromate furnishes 3 atoms of available oxygen, or 294 grm. of the salt yield 48 grm. of available oxygen, from which the equivalent weight

$$= \frac{294 \times 8}{48} = 49.$$

It will be seen from the equation that chrome alum can be obtained by the reduction of potassium bichromate in presence of sulphuric acid.

CHAPTER XII

MOLECULAR AND ATOMIC WEIGHT
DETERMINATIONS

THE fundamental ideas of equivalent, atomic, and molecular weights have been explained in Chapter II. As the atomic weight of an element is there stated to be the least weight of it (in grammes) ever present in one grammemolecule of any of its compounds, it is more convenient to discuss methods of molecular weight determination in the first instance. Analytical methods are employed to arrive at combining weights, i.e. equivalent weights, but additional data must be obtained in order to establish atomic weights. Cannizzaro (1860) pointed out that no consistent system of chemistry, and therefore no logical system of atomic weights, could be realized without the acceptance of Avogadro's hypothesis. Since molecular weights are proportional to gaseous densities, the fundamental method of arriving at them is by the determination of vapour density. Other methods of molecular weight determination, such as those dependent upon measurement of osmotic effects, are really subordinate to the former, and rest upon the application of Avogadro's reasoning to dilute solutions.

Methods of Determining Vapour Density. It has already been explained that the vapour density of a gas is its specific gravity with hydrogen as the standard. The vapour density of a gas such as oxygen or nitrogen is usually determined by direct weighing. A large glass globe fitted with a light capillary stopcock is weighed, firstly when exhausted as completely as possible, secondly when filled with hydrogen, and thirdly when filled with the gas under examination, the temperature and pressure being carefully noted. This gives the vapour density directly. Since only small differences in weight will be observed, many precautions are necessary to secure accuracy. The errors due to buoyancy and to adsorption of moisture on a glass surface are eliminated by counterpoising with a similar globe. The error due to shrinkage of the glass when a globe is evacuated has also to be corrected. The globe is weighed when filled with air at a known temperature and pressure.

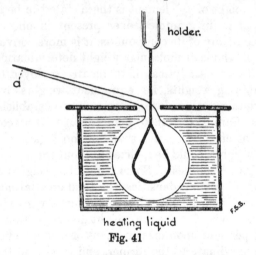

heating liquid

Fig. 41

The volume is determined by weighing it filled with water. These being known, the weight of the vacuous globe can be calculated, and also the weight of hydrogen it would contain.

A modification of this method was used by Dumas for determining the vapour densities of volatile solids and liquids. The apparatus is a glass (or porcelain, for very high temperatures) bulb (Fig. 41) of from 100 to 300 c.c. capacity, having a neck a of small diameter which can be drawn out to a fine capillary. The bulb is first weighed filled with air (w_1), the temperature (t_1) and pressure (p_1) being noted. Excess of the substance under examination is then introduced, and the bulb plunged into a bath of a suitable material (water, oil, molten lead, etc.) whose temperature is kept constant, and considerably above the boiling point of the substance. As the latter vaporizes the air is driven out of the bulb. When no more vapour issues, and this can be seen by holding a small flame at the end of the tube, the latter is sealed with a blowpipe, cleaned, and weighed (w_2): the temperature (t_2) being that of the bath, and the pressure (p_2) barometric. The volume of the bulb is found by opening the neck under water, which fills it almost completely. A residual bubble of air may be neglected if small, since the only error it can introduce is that caused by its change of volume with change of temperature. The bulb filled with water is weighed (w_3).

The vapour density is calculated from the various measurements as follows:

$w_3 - w_1 =$ weight of water (in grammes) = *effective* volume of the bulb (in c.c.).

       = volume of vapour at $t_2°$ C. and p_2 mm. of mercury pressure.

Now w_1 — weight of the vacuous bulb = weight of V c.c. of air at $t_1°$ C. and p_1 mm.

Since 1 c.c. of hydrogen at 0° C. and 760 mm. pressure weighs 0·00009 grm., and since the density of air is 14·4 times that of hydrogen, it is evident that the weight of V c.c. of air at $t_1°$ C. and p_1 mm. pressure will be

$$\frac{V \times p_1 \times 273}{760 \times (273 + t_1)} \times 0\cdot00009 \times 14\cdot4 \text{ grm.} = W_1 \text{ grm.}$$

Again, since the vacuous bulb weighs $w_1 - W_1$ grm., the weight of the vapour must be equal to $w_2 - (w_1 - W_1)$ grm. $= w_2 - w_1 + W_1$ grm.

The weight of V c.c. of hydrogen at $t_2°$ C. and p_2 mm. pressure can be calculated $= W_2$.

The vapour density of the substance, that is the ratio of the weight of the substance to that of an equal volume of hydrogen under the same conditions of temperature and pressure, is therefore given by the equation

$$\text{Vapour density} = \frac{w_2 - w_1 + W_1}{W_2}.$$

An actual determination of the vapour density of iodine by this method, conducted by Biltz in 1888, gave the following results:

     Weight of bulb filled with iodine vapour $= 25\cdot965$ grm.

     Weight of bulb filled with air $= 25\cdot458$ grm.

     Barometric pressure $= 743\cdot5$ mm.

     Temperature of room, $15\cdot6°$ C.

     Temperature of bath, $518°$ C.

     Effective capacity of the bulb $= 224$ c.c.

224 c.c. of air at $15\cdot6°$ C. and $743\cdot5$ mm. pressure weigh

$$\frac{224 \times 743\cdot5 \times 273}{760 \times 288\cdot6} \times 0\cdot00009 \times 14\cdot4 \text{ grm.} = 0\cdot269 \text{ grm.}$$

     Weight of vacuous bulb $= (25\cdot458 - 0\cdot269)$ grm.

         $= 25\cdot189$ grm.

     Weight of iodine vapour $= (25\cdot965 - 25\cdot189)$ grm.

         $= 0\cdot776$ grm.

     Weight of 224 c.c. of hydrogen at $518°$ C. and $743\cdot5$ mm.

$$= \frac{224 \times 743\cdot5 \times 273}{760 \times (273 + 518)} \times 0\cdot00009 \text{ grm.} = 0\cdot00681 \text{ grm.}$$

$$\text{Vapour density} = \frac{0\cdot776}{0\cdot00681} = 114.$$

For the diatomic molecule, I_2, the value should be 127, and for the monatomic molecule, I, $63\cdot5$. The result would appear to shew that at a temperature of $518°$ C. the dissociation of diatomic into monatomic molecules has taken place to an appreciable extent.

An apparatus in more general use for the determination of vapour densities is that devised by Victor Meyer, shewn in Fig. 42. The inner tube is carefully dried before being placed in position in the outer jacket, which contains a liquid capable of being heated well above the boiling point of the substance under examination. The latter is weighed in a small bulb which can be introduced into the inner tube on removal of the cork, dry sand or asbestos being placed at the bottom to avoid fracture when the little tube is dropped in. The bath is heated until the issue of bubbles from the delivery tube ceases. Its temperature is immaterial

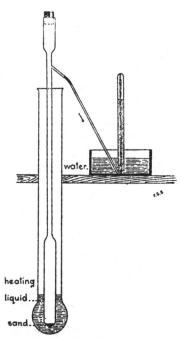

Fig. 42

provided, firstly, it is well above the boiling point of the substance to be examined, and secondly, that it shall remain constant during the short time required for the experiment. When the issue of air from the delivery tube has ceased, shewing that the temperature is steady, the cork is removed for an instant while the weighed tube is introduced, and the water-filled collecting tube placed over the delivery tube. As the substance is vaporized rapidly, it expels into the collecting tube a volume of air which is equal to the volume which the vapour would occupy if it could exist in the gaseous state at the temperature and pressure of the room. The temperature to be noted is *not* the temperature of the heating liquid but that of the room, since the air in the measuring tube has the same volume as the air replaced by the vapour, when cooled from the temperature of the bath to that of the room. The end of the experiment is reached when no more gas is expelled.

The collecting tube is removed to a deep vessel of water where the level can be adjusted and the volume of air read. The latter is calculated to normal temperature and pressure, and the weight of an equal volume of dry hydrogen deduced. The vapour density is then:

$$\frac{\text{weight of the substance taken}}{\text{weight of hydrogen calculated}}.$$

A determination of the vapour density of chloroform, with an outer bath of boiling water, gave the following numbers:

0·181 grm. of chloroform displaced 37·5 c.c. of air collected over water at 16° C. and 740 mm. pressure.

Pressure of aqueous vapour at 16° C. = 14 mm.

Corresponding volume of dry gas at 0° C. and 760 mm.

$$= \frac{37·5 \times 273 \times (740 - 14)}{(273 + 16) \times 760} = 33·8 \text{ c.c.}$$

$$\text{Vapour density} = \frac{0·181}{33·8 \times 0·00009} = 59·5.$$

Molecular weight = 119.

Victor Meyer's method of determining vapour density has been employed by Dewar, and in more recent years by

Nernst. The latter used iridium apparatus heated electrically, and determined the vapour densities of compounds such as the chlorides of potassium and sodium at 2000° C.

With the exception of the elements of the argon group, which form no compounds, atomic weight determinations are all in the first instance determinations of equivalent weights. The atomic weight of an element is always either identical with the equivalent weight, or is some very simple multiple of this, which is determined by methods described below. The meaning of hydrogen and oxygen equivalents is given in Chapter II. Any chemical change which directly or indirectly brings into action 1·008 parts by weight of hydrogen, or 8 parts by weight of oxygen, may *theoretically* be utilized for the purpose of determining the combining weight of an element, but substances differ greatly in suitability for this purpose. The choice of substances for careful determinations turns upon the possibility of preparing and maintaining them in pure condition. The methods chosen are those involving the minimum risk of errors in measurement.

In the case of such elements as sodium and potassium, for instance, the metal itself does not offer a suitable starting point for accurate determinations. The chemical activity of such metals militates against obtaining them in a state of purity, or of retaining them in that condition when so prepared. The chlorides, however, are remarkably stable compounds which can easily be prepared in a high state of purity; which do not tend to change under ordinary atmospheric conditions, and which can be weighed with great accuracy. By the interaction of these compounds with excess of silver nitrate, another stable and permanent compound, silver chloride, is produced, which can be collected and weighed without loss.

Practical considerations have led to the choice of oxygen

as the modern standard for atomic weights, its value being arbitrarily fixed at 16, and this element is the *primary* basis. Although the conversion of an element into its oxide is as a rule easily effected, determinations made in this way do not usually attain the highest accuracy. Since certain silver compounds are not only very easily prepared but are remarkably stable and definite, silver is largely used as a *secondary* basis. The ratio of silver to oxygen is therefore of fundamental importance, and is now very accurately known. One method of its determination, which also includes that of the equivalents of potassium and chlorine, is a classic of scientific research which should be studied in detail. In experimental work of this kind the materials are usually prepared by several independent methods, and processes of purification applied until their continuance no longer affects the result.

Berzelius, 1826, heated potassium chlorate until the weight of the residue was constant, and so determined the percentage of potassium chloride which remained. As a result of four concordant experiments, he found this to be 60·851.

Penny, 1839, employed a different method for decomposing the salt. Potassium chlorate was repeatedly evaporated with excess of concentrated hydrochloric acid until pure chloride remained, which was carefully dried and weighed.

$$KClO_3 + 6HCl = KCl + 3H_2O + 3Cl_2.$$

His apparatus consisted of two round-bottomed flasks placed in a horizontal position with the neck of one entering that of the other. A known weight of potassium chlorate was placed in the first flask, and evaporated with successive quantities of hydrochloric acid until constant weight was obtained. The liquid which distilled into the second flask was likewise evaporated to dryness and weighed. The object of the second flask was to retain any particles of solid accidentally carried over from the first flask. In Penny's hands this simple apparatus gave very satisfactory results. Six determinations gave a mean value of 60·8225 per cent. of potassium chloride obtained from potassium chlorate.

Stas, 1860, employing Berzelius's method with additional refinements, found the percentage yield of potassium chloride to be 60·8428. Using Penny's method of decomposing the chlorate with hydrochloric acid, a percentage of 60·849 was obtained.

These numbers represent the ratio potassium chloride/ oxygen, in which the extreme differences do not amount to one part in two thousand. Another stage was the determination of the ratio silver/potassium chloride.

Stas devoted much care to the preparation of pure silver, and some of his methods are to-day regarded as capable of yielding the metal in the highest state of purity obtainable. In one process the crude metal was dissolved in nitric acid, and precipitated as silver chloride. This, after separation and thorough washing, was reduced to the metallic state by warming with a solution of grape sugar in presence of caustic soda. The finely divided metal, again washed, was fused under a flux of borax and nitre and cast into ingots, or granulated by pouring while still molten into water. In some experiments the silver was actually distilled under the oxy-hydrogen flame in an apparatus of lime, but apparently no advantage accrued from this operation.

Stas's procedure was, briefly, as follows. A known weight of the metal was dissolved in nitric acid, and a known weight of potassium chloride, very nearly sufficient to react quantitatively with it, was dissolved in water and the solutions mixed. The silver halides possess the property of coagulating when the liquid containing them is thoroughly shaken. By titrating the supernatant liquid with dilute solutions ($N/100$ or less) of silver nitrate or of potassium chloride, according as the halide or silver solution was present in excess, the quantities of the two reacting materials were estimated with great precision. As the halides of silver are sensitive to actinic light, the experiments were conducted in a box illuminated by yellow light. The mean value of Stas's numbers gives the ratio silver/ potassium chloride = 100/69·103. The next stage was the determination of the ratio silver/silver chloride by the direct synthesis of silver chloride from pure silver. This was carried out both by heating the metal in excess of chlorine to constant weight, and also by dissolving the pure metal in nitric acid and precipitating the chloride, which was collected and weighed as such. The mean values obtained were, silver/silver chloride = 100/132·8445, and therefore silver/chlorine = 100/32·8445.

By neglecting small fractions from the values given above, and accepting the formula $KClO_3$ for potassium chlorate, from the decomposition of this salt (100 − 60·85) grm. of oxygen correspond to 60·85 grm. of potassium chloride, and therefore 48 grm. of oxygen correspond to

$$\frac{60·85 \times 48}{(100 - 60·85)} \text{ grm. of potassium chloride, } = 74·6,$$

which is, therefore, the formula weight of potassium chloride.

Again, as 69·1 grm. of potassium chloride react with 100 grm. of silver, 74·6 grm. of the salt react with

$$\frac{100 \times 74·6}{69·1} \text{ grm. of silver, } = 108, \text{ which is the atomic weight}$$

of silver.

Further, since 100 grm. of silver react with 32·8 grm. of chlorine, 108 grm. of the metal combine with

$$\frac{32·8 \times 108}{100} = 35·46 \text{ grm.}$$

of chlorine, and therefore the atomic weight of chlorine = 35·46.

Lastly, by subtracting the atomic weight of chlorine from the formula weight of potassium chloride, the atomic weight of potassium is obtained, = 39·14.

The atomic weight of carbon has been determined by burning the element to carbon dioxide, and weighing the latter after absorption in caustic potash. The impurities in the diamond or graphite used were very small in amount, and could be allowed for by weighing the residual ash. The products of combustion were passed over red hot copper

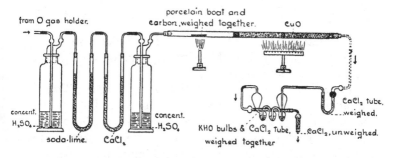

Fig. 43

oxide, to ensure the oxidation to the dioxide of any carbon monoxide formed. The details of the experiment will be made clear by a reference to Fig. 43.

Better results were obtained by Stas, who oxidized carbon monoxide to the dioxide by passage of the carefully purified gas over a known weight of heated copper oxide. The weights of the residual copper and of the carbon dioxide formed were determined, the principle of the experiment being similar to that employed by Dumas in his work on the gravimetric composition of water. Stas found that 1 grm. of oxygen produced 1·750 grm. of carbon dioxide, from which the atomic weight of carbon is very nearly 12.

The atomic weight of nitrogen has been the subject of numerous investigations. Many methods have been employed, but the direct conversion of silver into silver nitrate is perhaps the best. This was used with the following results.

Weight of silver	Weight of $AgNO_3$	Observer	Date
100 grm.	157·441 grm.	Penny	1839
100 ,,	157·484 ,,	Stas	1865
100 ,,	157·479 ,,	Richards and Forbes	1907

Calculating from the known atomic weights of silver and oxygen, the atomic weight of nitrogen given by these remarkably concordant results is 14·01.

Deduction of Atomic Weights from Equivalent Weights. Equivalent weight determinations are simply measurements of combining ratios, and require to be supplemented by additional data in order to ascertain the atomic weight. The following are the more important considerations.

Obviously one gramme-molecular weight of a compound must contain not less than one gramme-atomic weight of any element in it. For instance, the equivalent weight of oxygen is 8. As the density of steam is 9, its molecular weight is 18. Therefore the atomic weight of oxygen must be either 8 or 16; it clearly cannot be *greater* than 16 because of the limit set by the molecular weight of steam.

The *inferior* limit of the atomic weight of an element is the *maximum* equivalent weight, while the *superior* limit is obtained by the examination of a sufficiently large number of compounds. No oxygen compound has ever been found to contain *less* than 16 grm. of oxygen per

gramme-molecule, and hence the atomic weight is settled
by experimental evidence at 16. It will be noted that this
is not *proof* that the true atomic weight has been obtained,
but only probability. If the number of compounds examined be large, it may be probability of a high order
however, and the conclusions reached in this way are in
general confirmed by one or more of the considerations
which follow. Again, the maximum equivalent weight of
phosphorus is 10·35. The vapour density of phosphine is
17, and the molecular weight 34. The weight of phosphorus
in phosphine is approximately 31 grm. per gramme-
molecule. This gives the value of the maximum atomic
weight of phosphorus as $10·35 \times 3 = 31·05$, and no smaller
value than this has ever been found in any phosphorus
compound.

Dulong and Petit in 1819 pointed out that when the
specific heat of a solid element is multiplied by the accepted atomic weight, a roughly constant value of about
6·3 is obtained. This they called the atomic heat, and concluded that the atomic heats of all solid elements have this
value, a conclusion approximately true for such elements
as happened to have assigned to them at that time the
values now believed to be the correct multiples of their
equivalent weights.

The 'law' of Dulong and Petit can therefore be stated:

Atomic weight × specific heat = 6·3 (*atomic heat*).

The variations in the case of many of the elements are
however very considerable, and the equation 6·3/specific
heat = atomic weight, gives only a rough approximation
to the true atomic weight. Since, however, the atomic
weight = equivalent weight × n, where n is a small whole
number, even wide discrepancies still allow the value of n
to be fixed from measurements of specific heat. As an
example, the equivalent weight of calcium is 20·03, and

therefore its true atomic weight $= 20.03 \times n$. Applying the law, the specific heat of calcium being 0.15:

$$\text{Approximate atomic weight} = \frac{6.3}{0.15} = 42.$$

From this, $n = 2$, and the true atomic weight becomes $20.03 \times 2 = 40.06$.

Although in some cases the variations in the specific heat (according to the temperature at which it is taken) are small, and in practice negligible, this is by no means always true. The specific heat of diamond, for instance, varies from 0.113 at $10°$ C. to 0.459 at $1000°$ C., that is to say, it increases more than fourfold. The temperature at which the specific heat is determined is therefore all important. It has been found that in such cases the best agreement with the law is obtained by taking the specific heat at as high a temperature as possible, short of fusion.

The importance of the 'law' of Dulong and Petit has receded in recent years, since other and more reliable methods of fixing atomic weights are known. Furthermore, the variations in the law are so wide as to inspire distrust, particularly as no acceptable explanation of the discrepancies has yet been offered.

It has long been observed that certain salts crystallize in particular geometric forms, and different salts which crystallize in the same form are said to be isomorphous. In a restricted sense, isomorphism demands the formation of mixed crystals and layer crystals (Ostwald). In 1821 Mitscherlich pointed out that isomorphous salts are built up in a similar way; that is, the component elements in the one are equal in valency to the corresponding elements in the other. Thus chromates are isomorphous with the corresponding sulphates. This principle has been applied to the fixing of atomic weights. An interesting historic example concerns the rare element gallium, discovered by Lecoq de Boisbaudran in 1875. The oxygen equivalent of this metal was found to be 23·3. Its sulphate forms a double salt with ammonium sulphate, isomorphous with the alums, which therefore is also an alum. Since aluminium is tervalent, gallium is also tervalent, and the atomic weight becomes 23·3 × 3 = 69·9.

The system of classification of the elements known as the Periodic Law depends upon the relationship between certain elements, as shewn by their properties and those of their compounds. The law is usually stated thus. *The properties of the elements are a periodic function of their atomic weights.* This has proved a valuable means of controlling atomic weights, but it can only be considered with advantage when dealing with a greater number of the elements than the present work contemplates, and the reader is therefore referred to larger treatises.

Isotopes. As early as 1815 Prout, observing the close approximation to whole numbers in the atomic weights of many of the elements, advanced a hypothesis which could be interpreted to mean that the atom of every element was an agglomeration of hydrogen atoms; that there was, in fact, but one kind of ultimate atom in nature, the hydrogen atom. Since investigation shewed conclusively that the fractional atomic weight of chlorine, for instance, could not be attributed to experimental error, the unit was later assumed as half an atom of hydrogen. That is, that the hydrogen atom itself was assumed to be made up of two atoms of some (unknown) substance. Even this, however, could not be maintained in face of the accuracy with which the atomic weights of certain elements had been determined, and which were certainly not simple multiples of half a unit. Further, this kind of subdivision might be continued indefinitely, and this destroyed all interest in the hypothesis, which soon fell into obscurity. It was never quite abandoned however, and recent investigations have recalled the hypothesis, and done much to re-establish it on a new basis. Such an element as chlorine, with a fractional atomic weight of 35·46, is now known to consist of two kinds of chlorine atom, both chemically alike as far as is known, but differing in weight. There are, for instance, atoms with an atomic weight of 35, others with an atomic weight of 37, and possibly still others, and all these *isotopes* have *integral* atomic weights. Whatever the method of preparation may be, the pure gas is found experimentally always to have the same density, and it follows that the isotopes are invariably present in the same proportions. When the atomic weight of chlorine is determined by *chemical* methods, the result is the average weight of an enormous number of atoms, and this average value is 35·46. Very many of the elements are now known to have isotopes, but the difficulties of isolating them in other than mere traces are very great, and no satisfactory results have hitherto been obtained. The discovery of isotopes has diminished the theoretical importance of atomic weights, but has in no way lessened their practical utility.

International Atomic Weights.

	Symbol	Atomic weight		Symbol	Atomic weight
Aluminium ...	Al	27·1	Neodymium ...	Nd	144·3
Antimony ...	Sb	120·2	Neon	Ne	20·2
Argon	A	39·9	Nickel	Ni	58·68
Arsenic	As	74·96	Niton (radium		
Barium	Ba	137·37	emanation) ...	Nt	222·4
Bismuth... ...	Bi	208·0	Nitrogen ...	N	14·008
Boron	B	10·9	Osmium... ...	Os	190·9
Bromine... ...	Br	79·92	Oxygen	O	16·00
Cadmium ...	Cd	112·40	Palladium ...	Pd	106·7
Caesium... ...	Cs	132·81	Phosphorus ...	P	31·04
Calcium	Ca	40·07	Platinum ...	Pt	195·2
Carbon	C	12·005	Potassium ...	K	39·10
Cerium	Ce	140·25	Praseodymium	Pr	140·9
Chlorine... ...	Cl	35·46	Radium	Ra	226·0
Chromium ...	Cr	52·0	Rhodium ...	Rh	102·9
Cobalt	Co	58·97	Rubidium ...	Rb	85·45
Columbium ...	Cb	93·1	Ruthenium ...	Ru	101·7
Copper	Cu	63·57	Samarium ...	Sa	150·4
Dysprosium ...	Dy	162·5	Scandium ...	Sc	45·1
Erbium	Er	167·7	Selenium ...	Se	79·2
Europium ...	Eu	152·0	Silicon	Si	28·3
Fluorine... ...	F	19·0	Silver	Ag	107·88
Gadolinium ...	Gd	157·3	Sodium	Na	23·00
Gallium	Ga	70·1	Strontium ...	Sr	87·63
Germanium ...	Ge	72·5	Sulphur	S	32·06
Glucinum ...	Gl	9·1	Tantalum ...	Ta	181·5
Gold	Au	197·2	Tellurium ...	Te	127·5
Helium	He	4·00	Terbium... ...	Tb	159·2
Holmium ...	Ho	163·5	Thallium ...	Tl	204·0
Hydrogen ...	H	1·008	Thorium... ...	Th	232·15
Indium	In	114·8	Thulium... ...	Tm	168·5
Iodine	I	126·92	Tin	Sn	118·7
Iridium	Ir	193·1	Titanium ...	Ti	48·1
Iron	Fe	55·84	Tungsten ...	W	184·0
Krypton... ...	Kr	82·92	Uranium ...	U	238·2
Lanthanum ...	La	139·0	Vanadium ...	V	51·0
Lead	Pb	207·20	Xenon	Xe	130·2
Lithium	Li	6·94	Ytterbium		
Lutecium ...	Lu	175·0	(Neoytterbium)	Yb	173·5
Magnesium ...	Mg	24·32	Yttrium... ...	Yt	89·33
Manganese ...	Mn	54·93	Zinc	Zn	65·37
Mercury... ...	Hg	200·6	Zirconium ...	Zr	90·6
Molybdenum ...	Mo	96·0			

INDEX OF AUTHORS

INDEX OF SUBJECTS

Printed in the United States
By Bookmasters